SHOW WHAT YOU KNOW® ON THE 9TH GRADE

FCAT

FLORIDA COMPREHENSIVE ASSESSMENT TEST

MATHEMATICS

grade
9

TEST-PREPARATION FOR THE
FLORIDA COMPREHENSIVE
ASSESSEMENT TEST

Show What You Know®
Publishing

Published by:

Show What You Know® Publishing
A Division of Englefield & Associates, Inc.
P.O. Box 341348
Columbus, OH 43234-1348
Phone: 1-877-PASSING (727-7464)
www.showwhatyouknowpublishing.com
www.passthefcat.com

FCAT Item Distribution information was obtained from the Florida Department of Education Website, January 2006.

Printed in the United States of America
08 07 06   20 19 18 17 16 15 14 13 12 11 10 9 8 7 6 5 4 3 2 1

ISBN: 1-59230-162-2

Acknowledgements

Show What You Know® Publishing acknowledges the following for their efforts in making this assessment material available for Florida students, parents, and teachers.

Cindi Englefield, President/Publisher
Eloise Boehm-Sasala, Vice President/Managing Editor
Christine Filippetti, Project Editor
Jill Borish, Project Editor
Erin McDonald, Project Editor
Heather Holliday, Project Editor
Charles V. Jackson, Project Editor
Jennifer Harney, Illustrator/Cover Designer

About the Contributors

The content of this book was written BY teachers FOR teachers and students and was designed specifically for the Florida Comprehensive Assessment Test (FCAT) for Grade 9. Contributions to the Mathematics section of this book were also made by the educational publishing staff at Show What You Know® Publishing. Dr. Jolie S. Brams, a clinical child and family psychologist, is the contributing author of the Test Anxiety and Test-Taking Strategies chapters of this book. Without the contributions of these people, this book would not be possible.

Table of Contents

Introduction

Dear Student:

This *Show What You Know® on the FCAT for Grade 9 Mathematics, Student Self-Study Workbook* was created to give you practice in preparation for the Florida Comprehensive Assessment Test (FCAT) in Mathematics.

The first two chapters in this workbook—Test Anxiety and Test-Taking Strategies—were written especially for ninth-grade students. Test Anxiety offers advice on how to overcome nervous feelings you may have about tests. The Test-Taking Strategies chapter includes helpful tips on how to answer questions correctly so you can succeed on the Mathematics section of the FCAT.

In the Mathematics chapter of this book, you will find the following:
- A Mathematics Glossary, a Glossary of Illustrations, and a Mathematics Reference Sheet is provided to review mathematics terms.
- Scoring Guides for multiple-choice and gridded-response questions.
- After each mathematics question in the tutorial, you will answer multiple-choice or gridded-response questions. An analysis for each tutorial question is given to help you identify the correct answer.
- Two full-length Mathematics Assessments follow the tutorial section for additional Mathematics practice. An Answer Key will help you check to see if you answered the assessment questions correctly.

This Student Self-Study Workbook also includes a Mathematics Assessment Correlation Chart. This chart can be used to identify individual areas of needed improvement.

This book will help you become familiar with the look and feel of the FCAT Mathematics Assessment and will provide you with a chance to practice your test-taking skills to show what you know.

Good luck on the FCAT!

Test Anxiety

What Is Test Anxiety?

Test anxiety is a fancy term for feelings of worry and uneasiness that students feel before or during a test. Almost everyone experiences some anxiety at one time or another. Experiencing feelings of anxiety before any challenge is a normal part of life. However, when worrying about tests becomes so intense it interferes with test taking, or if worrying causes students mental or physical distress, this is called test anxiety.

What Are the Signs of Test Anxiety?

Test anxiety is much more than feeling nervous. In fact, students will notice test anxiety in four different areas: thoughts, feelings, behaviors, and physical symptoms. No wonder test anxiety gets in the way of students doing or feeling well.

1. Thoughts

Students with test anxiety usually feel overwhelmed with negative thoughts about tests and about themselves. These thoughts interfere with the ability to study and to take tests. Usually, these bothersome thoughts fall into three categories:

- **Worrying about performance**—A student who worries may have thoughts such as, "I don't know anything. What's the matter with me? I should have studied more. My mind is blank; now I'll never get the answer. I can't remember a thing; this always happens to me. I knew this stuff yesterday and now I can't do anything."

- **Comparing oneself to others**—A student who compares performance might say, "I know everyone does better than I do. I'm going to be the last one to finish this. Why does everything come easier for everyone else? I don't know why I have to be different than others."

- **Thinking about possible negative consequences**—A student with negative thoughts would think, "If I don't do well on this test, my classmates will make fun of me. If I don't do well on the FCAT Mathematics test, my guidance counselor will think less of me. I won't be able to go to my favorite college. My parents are going to be angry."

Many of us worry or have negative thoughts from time to time. However, students with test anxiety have no escape and feel this worry whenever they study or take tests.

2. Feelings

In addition to having negative thoughts, students with test anxiety are buried by negative feelings. Students with test anxiety often feel:

- **Nervous and anxious**—Students feel jittery or jumpy. Anxious feelings may not only disrupt test taking but may interfere with a student's life in other ways. Small obstacles, such as misplacing a book, forgetting an assignment, or having a mild disagreement with a friend, may easily upset students. They may become preoccupied with fear, may have poor self-esteem, and may feel that the weight of the world is on their shoulders. They seem to be waiting for "the next bad thing to happen."

- **Confused and unfocused**—Students with test anxiety have their minds in hundreds of anxious places. They find it difficult to focus on their work, which makes studying for tests even harder. Students with test anxiety also have difficulty concentrating in other areas. When they should be listening in class, their minds worry about poor grades and test scores. They jump to conclusions about the difficulty of an upcoming test. They find themselves fidgeting. They constantly interrupt themselves while studying, or they forget how to complete simple assignments. Anxiety can interfere with a student's ability to focus, study, and learn.

- **Angry and resentful**—Test anxiety can lead to irritable and angry feelings. Anxious students are defensive when communicating with others. They become overwhelmed by negative thoughts and feel they are not good enough. Test anxiety also makes students feel "trapped" and as though they have no escape from school or tests. Students who feel there is no way out may get angry; they may resent the situation. They feel jealous of people they believe find school easier. They are angry at the demands placed on them. The more angry and resentful students become, the more isolated and alone they feel. This only leads to further anxiety and increased difficulties in their lives.

• **Depressed**—Anxiety and stress can lead to depression. Depression sometimes comes from "learned helplessness." When people feel they can never reach a goal and that they are never good enough to do anything, they tend to give up. Students who are overly anxious may get depressed. They lose interest in activities because they feel preoccupied with their worries about tests and school. It might seem as though they have no time or energy for anything. Some students with test anxiety give up on themselves completely, believing if they cannot do well in school (even though this may not be true), then why bother with anything?

Not all students with test anxiety have these feelings. If you or anyone you know seem to be overwhelmed by school, feel negative most of the time, or feel hopeless about school work (test taking included), you should look to a responsible adult for some guidance.

3. Behavior

Students with test anxiety often engage in behavior that gets in the way of doing well. When students have negative thoughts and feelings about tests, they participate in counterproductive behavior. In other words, they do things that are the opposite of helpful. Some students avoid tests altogether. Other students give up. Other students become rude and sarcastic, making fun of school, tests, and anything to do with learning. This is their way of saying, "We don't care." The truth is, they feel anxious and frustrated. Their negative behaviors are the result of thoughts and feelings that get in the way of their studying and test taking.

4. Physical Symptoms

All types of anxiety, especially test anxiety, can lead to very uncomfortable physical symptoms. Thoughts control the ways in which our bodies react, and this is certainly true when students are worried about test taking. Students with test anxiety may experience the following physical symptoms at one time or another:

- sweaty palms
- stomach pains
- "butterflies" in the stomach
- difficulty breathing
- feelings of dizziness or nausea

- headaches
- dry mouth
- difficulty sleeping, especially before a test
- decrease or increase in appetite

Test anxiety causes real physical symptoms. These symptoms are not made up or only in your head. The mind and body work together when stressed, and students can develop uncomfortable physical problems when they are anxious, especially when facing a major challenge like the FCAT.

The Test Anxiety Cycle

Have you ever heard the statement "one thing leads to another?" Oftentimes, when we think of that statement, we imagine Event A causes Event B, which leads to Event C. For example, being rude to your younger brother leads to an argument, which leads to upset parents, which leads to some type of punishment, like grounding. Unfortunately, in life, especially regarding test anxiety, the situation is more complicated. Although one thing does lead to another, each part of the equation makes everything else worse, and the cycle just goes on and on.

Let's think back again to teasing your younger brother. You tease your younger brother and he gets upset. The two of you start arguing and your parents become involved. Eventually, you get grounded. Sounds simple? It might get more complicated. When you are grounded, you might become irritable and angry. This causes you to tease your little brother more. He tells your parents, and you are punished again. This makes you even angrier, and now you don't just tease your little brother, you hide his favorite toy. This really angers your parents who now do not let you go to a school activity. That upsets you so much you leave the house and create trouble for yourself. One thing feeds the next. Well, the same pattern happens in the test anxiety cycle. Look at the following diagram.

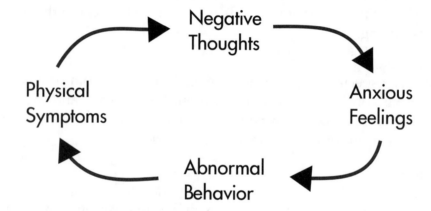

As you can see, the four parts of this diagram include the thoughts, feelings, behavior, and physical symptoms discussed earlier. When a student has test anxiety, each area makes the others worse. The cycle continues on and on. Here's an example:

Let's start off with some symptoms of negative thinking. Some students might say to themselves, "I'll never be able to pass the FCAT Mathematics test!" This leads to feelings of frustration and anxiety. Because the student has these negative thoughts and feelings, his or her behavior changes. The student avoids tests and studying because they are nerve racking. Physical symptoms develop, such as the heart racing or the palms sweating. Negative thoughts then continue, "Look how terrible I feel; this is more proof I can't do well." The student becomes more irritable, even depressed. This affects behavioral symptoms again, making the student either more likely to avoid tests or perhaps not care about tests. The cycle goes on and on and on.

Is Test Anxiety Ever Good?

Believe it or not, a little worrying can go a long way! Too much test anxiety gets in the way of doing one's best, but students with no anxiety may also do poorly. Studies have shown that an average amount of anxiety can help people focus on tasks and challenges. This focus helps them use their skills when needed. Think about a sporting event. Whether a coach is preparing an individual ice skater for a competition or is preparing the football team for the Friday night game, getting each athlete "psyched up" can lead to a successful performance. A coach or trainer does not want to overwhelm the athlete. However, the coach wants to sharpen the senses and encourage energetic feelings and positive motivation. Some schools have a team dinner the night before a competition. This dinner provides some pleasant entertainment, but it also focuses everyone on the responsibilities they will have the next day.

Consider the graph below. You can see that too little test anxiety does not result in good test scores. As students become more concerned about tests, they tend to do better. But wait! What happens when too much anxiety is put into the equation? At that point, student performance decreases remarkably. When anxiety reaches a peak, students become frustrated and flustered. Their minds tend to blank out, they develop physical symptoms, they cannot focus, and they also behave in ways that interfere with their performance on tests.

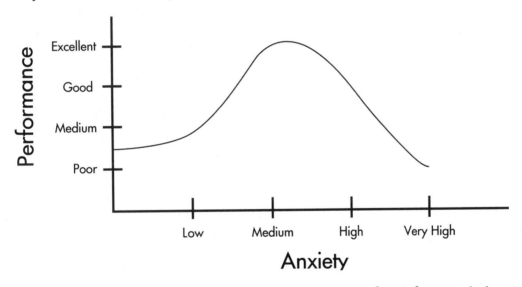

An important key to successful test taking is to get yourself in the right mood about taking a test. Looking at a test as a challenge and looking forward to meeting that challenge, regardless of the end result, is a positive and healthy attitude. You will feel excited, motivated, and maybe a little nervous but certainly ready to face the FCAT.

How Do I Tackle Test Anxiety?

Although test anxiety is an uncomfortable and frustrating feeling, the good news is you can win the battle over test anxiety! Conquering test anxiety will not be accomplished by luck or magic, but it can be done by students of all ages in a relatively short period of time. If you can learn to master test anxiety at this point in your life, you will be on the road to successfully facing many other challenges you will encounter.

1. Change the Way You Think

Whether you realize it or not, your thoughts—good and bad—influence your life. The way we think is related to how we feel about ourselves, how we get along with other people, and how well we do in school, especially when taking tests.

- *Positive Thinking Can Block Out Negative Thinking*—It is impossible to think two opposite thoughts at the same time. You may have one idea and then think about another, but one is always going to "win" over the other. When you practice positive thinking, you are replacing negative thoughts with positive ones. The more you are able to think positive thoughts, the less you will be troubled by negative ones.

- *The Soda Pop Test*—It's just as easy to have positive thoughts as negative ones. Everyone has heard the saying, "There is more than one side to any story." Just as there are two opinions on any given subject, there is generally more than one way to look at almost every situation in life. Some ways are more helpful than others.

Think about a can of soda pop. Draw a line down the middle of a blank piece of paper. On one side, put the heading, "All the bad things about this can of soda pop." On the other side, put another heading, "All the good things about this can of soda pop." Now, write appropriate descriptions or comments under each heading. For example, you could write, "This can of soda pop is a lot smaller than a two-liter bottle," which is negative thinking. Or, you could write, "This can of soda pop is just the right size to stay cold and fizzy until I finish it." It's easy to look at the soda-pop can and think bad thoughts. But you are also able to come up with many good things. If you spent all your time focusing on the negative aspects, you might believe the can of soda pop is bad. It is better to look at the positive side of things.

Part of successful test taking has to do with how you look at tests. With the can of soda pop, you could choose to think negatively, or you could have positive thoughts. The same holds true for tests. You can look at a test as a scary or miserable experience, or you can look at a test as just one of many challenges you will face in your life.

Counselors have known for years that people who are worried or anxious can become happier when thinking positive thoughts. Even when situations are scary, such as going to the dentist or having a medical test, "positive imagery" is very helpful. Positive imagery simply means focusing on good thoughts to replace anxious thoughts.

You can replace negative thoughts with positive ones through practice. Believe it or not, it really works!

- *Thoughts of Success*—Thinking "I can do it" thoughts chases away ideas of failure. Times that you were successful, such as when you did well in a sports event or figured out a complicated math question, are good things to think about. Telling yourself you have been successful in the past and can now master the Mathematics section of the FCAT will replace thoughts that might otherwise cause anxiety.

- *Relaxing Thoughts*—Some people find that thinking calming or relaxing thoughts is helpful. Picturing a time in which you felt comfortable and happy can lessen your anxious feelings. Imagining a time when you visited the ocean, climbed a tree, or attended a concert can help you distract your mind from negative thoughts and focus on times that you were relaxed, happy, and felt positive.

- *All-or-Nothing Thinking*—Nothing is ever as simple as it seems. Sometimes we convince ourselves something is going to be "awful" or "wonderful," but it rarely turns out that way.

No test is "completely awful" or "completely perfect." Tests are going to have easy questions and hard questions, and you are going to have good test days and bad test days. The more you set up expectations that are all positive or negative, the more stressful the situation becomes. Accepting that nothing is totally good or bad, fun or boring, or easy or hard will reduce your anxiety and help you set reasonable expectations about tests. When you think about tests, try not to think about them as the road to academic success or a pit of failure. Instead, realize that all challenges have both good and bad elements, and you have to take everything in stride.

- *Making "Should" Statements*—Making "should" statements sets students up for failure. Sure, it is important to try your best, to study hard, and to make a reasonable effort on the FCAT; it may even be good to take an extra study session, try another practice test, or ask a teacher or tutor for advice and suggestions. It is also a good idea to use a book such as this one to help you do your best and show what you know. However, there is a big difference between doing your "reasonable best" and living your life with constant worries and put-downs. Students who constantly tell themselves "I should" and berate themselves for not having done everything possible only increase their levels of anxiety.

Go back to the test anxiety cycle. Suppose your thoughts are, "I should have stayed up an extra hour and studied," or "I should have reviewed those geometry formulas." The more you think these thoughts, the more anxious you get. The more anxious you get, the worse you feel. Again, the cycle goes on and on.

One part of maturing is learning to balance your life. Life is happiest when you find a good balance between being a lazy do-nothing and being a perfectionist. While we all know laziness gets us nowhere, being a perfectionist may actually paralyze your future chances of success because you will eventually fear meeting any new challenges. Failure does not mean real failure; it just means being imperfect. Preventing perfectionism begins by saying "no" to unreasonable thoughts and "should" statements. "Should" statements place high demands on a student and only lead to frustration and feelings of failure, shame, and anxiety.

Students who always think about what they "should" do often exhaust themselves by doing too much and worrying excessively. Exhaustion is another factor that leads to poor test-taking results.

Breaking the "should" habit means replacing "should" statements with positive comments about what you have accomplished and what you hope to reasonably accomplish in the future. For example, instead of saying, "I shouldn't have gone to the football game," or "I should have stayed home and studied," say, "I studied for two hours before the football game, and then I had a good time. Two hours was plenty to study for a geometry quiz. I need to have time for friends as well as studying. I concentrated while studying, and I think I did a good job. Even if I don't get a perfect score on the geometry quiz, I know I will do pretty well, and I gave myself the opportunity to do my best."

2. Control Physical Symptoms

Changing your physical response to stress can help break the test anxiety cycle. Relaxing is difficult when facing a major challenge such as the FCAT, but there are many proven techniques that can help you calm down.

- *Relax the Morning of the Test*—Try to allow yourself to relax the morning of the test. Engaging in some mild exercise, such as taking a walk, will relieve a lot of your physical stress. Some students may find that a workout the night before an exam makes them feel more relaxed and helps them sleep well. This is probably because the exercise distracts the student from the upcoming test. Also, intense exercise releases chemicals in the brain that cause you to feel calmer and happier. It may only take a quick walk around the block to help you relax and get your mind off your problems.

- *Listen to Music*—Listening to music in the morning before a test may also be helpful for students. It probably doesn't matter what kind of music you listen to as long as it makes you feel good about yourself, confident, and relaxed.

- *Relaxation Exercises*—Relaxation exercises are helpful to many students. Stress causes many physical changes in the body, including tenseness in all muscle groups, increased heart rate, and other physical symptoms. Learning simple exercises to feel less tense can also help break the test anxiety cycle.

Most exercises include tightening and releasing tension in your body as well as deep breathing. The purpose of all of these exercises is to distract you from the anxiety of an upcoming test and to allow your body to feel more loose and relaxed. These exercises can be completed while sitting at your desk, taking a test, or studying.

Try this simple relaxation exercise the next time you are tense. Sit upright in your chair, but allow yourself to be comfortable. Close your eyes and take four deep breaths in and out. When you get to the fourth breath, start breathing quietly but remain focused on your breathing. Start increasing the tension in your feet by squeezing your toes together tightly and then slowly releasing the pressure. Feel how relaxed your toes are feeling? Now tighten and release other muscle groups. Go from your legs to your stomach, to your shoulders, to your hands, and finally to your forehead. Squeeze and tighten your muscles and then relax them, all while focusing on your breathing. Once you practice this strategy, you might be able to feel more relaxed in a matter of seconds. This would be a good strategy to use during tests when you feel yourself becoming unfocused and anxious.

Prepare For the FCAT Mathematics Assessment and Change the Way You Behave

Preparation always reduces anxiety. Taking the FCAT Mathematics test seriously, trying to do well on practice tests, and making an effort in all your classes will help you feel more confident and relaxed about the FCAT. Learning test-taking strategies can also give you a feeling of power and control over the test. No feeling is worse than realizing you are not prepared. Going into a test without ever having reviewed the FCAT Mathematics material, looked at test-taking strategies, or concentrated on your schoolwork is very much like jumping out of an airplane without a parachute. You would be foolish if you were not panicked. Looking at the FCAT Mathematics section as just one more reason to take school seriously will help your grades, attitude, and success on the test.

1. Use Mental Preparation
Before the test, imagine in step-by-step detail how you will perform well and obtain a positive result. Several days before the test, think through the day of the test; repeat this as many times as you need. Imagine getting up in the morning, taking a nice shower, getting dressed in comfortable clothes, and listening to music on your way to school. Think about sitting in the testing room with a confident expression on your face. Imagine yourself remembering all of the strategies you read about in this book and learned in your classroom. Go through an imaginary test, step by step, practicing what you will do if you encounter a difficult question. You should also repeat the positive thoughts that should go through your head during the test. Preparation like this is key for reducing anxiety, as you already feel you have taken the test prior to ever having stepped in that testing room!

2. Don't Feel Alone

People feel more anxious when they feel alone and separate from others. Have you ever worried about a problem in your family or something going wrong at school? Things seem much worse when you are alone, but when you talk to someone who cares about you, you will find your problems soon seem less worrisome. Talk to your friends, parents, and teachers about your feelings. You will be surprised at the support you receive. Everyone has anxious feelings about tests. Having others understand your anxious feelings will help you accept yourself even more. Other people in your life can also give you suggestions about tests and can also help you put the FCAT Mathematics and other tests in perspective.

3. Congratulate Yourself During the Test

Students with test anxiety spend a lot of time putting themselves down. They have never learned to say good things about themselves or to congratulate themselves on successes. As you go through the FCAT Mathematics, try to find ways to mentally pat yourself on the back. If you find yourself successfully completing a difficult question, tell yourself you did a good job. When you finish reading a Mathematics test item and feel you understand the information fairly well, remind yourself you are doing a good job in completing the FCAT Mathematics. Paying attention to your successes, and not focusing on your failures, can greatly reduce test anxiety.

Test-Taking Strategies

Understand the Types of Questions on the Mathematics Section of the FCAT

In preparing for the Mathematics section of the FCAT, you will need to think about the various types of questions you might be asked, but you also must think about and practice the different types of answers that will be required.

Imagine being five years old. It was your grandmother's birthday, so you drew her a picture of a birthday cake and a flower. It was cute, but that was the only way you could give her a gift. You didn't have any money, and you didn't have a lot of experience in making gifts or sharing birthday wishes. Now, since you're older, you have lots of choices as to how you can wish her "Happy Birthday."

As a high school student, you have been exposed to many ways to express a mathematical answer or concept, and the Mathematics section of the FCAT expects you to be able to show what you know in more than one way.

Use a Checklist to Think Through Problems

Do you remember elementary school? Remember your teacher giving you checklists to help you do your best? Well, checklists are still helpful tools, especially on the Mathematics FCAT.

Even though much of the mathematics that you are learning is more complicated than what you were taught as a child, having a set pattern of reviewing your work can lead to a better result on this test.

As you tackle mathematics questions, get in the habit of asking yourself these questions:

☑ Did I read the problem carefully? Misreading the problem will certainly lead to a poor result.

☑ What solution does the problem require? Does it ask for a single number, or for something else?

☑ Did I answer the problem in the manner it was asked?

☑ Did I use all the tools I have to answer the problem? Could a calculator have been helpful?

☑ Did I write legibly? Can I read my written numbers?

☑ Did I use mathematical vocabulary and language whenever I could?

☑ Does my answer make sense? Based on my knowledge of mathematics, is this a reasonable answer?

Know Your Tools

On the Mathematics section of the FCAT, you may use a calculator to find answers. These questions test your ability to understand the process of finding a correct answer, not your skills at quickly performing routine calculations without help. Unfortunately, using a calculator won't be helpful if you don't know how to quickly and accurately use a calculator, and may actually cause you to arrive at the wrong answer.

Using a calculator takes practice. When you first learned a computer keyboard, you probably were slower and made many more mistakes. If you rely on your parents or friends to type for you, you probably haven't improved your keyboarding skills very much since elementary school. If you had to type your answers on the FCAT, it would take you a long time, and you would have many mistakes. The same amount of practice is needed for a calculator in order for it to be helpful to you on the Mathematics section of the FCAT. Students who avoid using a calculator except when absolutely necessary will not benefit from using one when taking the FCAT. So, if you are not comfortable with a calculator, practice! It will not only help you on the test, but you will find this skill valuable in jobs and for solving problems in your everyday life.

Know Your Vocabulary

When most students think of vocabulary words, they think of Reading and Writing, not Mathematics. In fact, having a good understanding of mathematical words will be as helpful to you on the FCAT as having a comprehensive vocabulary in writing and speaking.

You have been taught many important mathematical terms since elementary school, and many of them will be in the questions and answers on the Mathematics section of the FCAT. A good strategy when preparing for this test is to make a mathematics vocabulary list with correct definitions and examples. Take a peek at some of your old textbooks and the ones you are using now. You could even go to the library and find some middle school textbooks or other mathematics books. Every day, find one or two words and get to know them well! In a short time, you will be surprised to find that you have mastered a fairly large number of mathematical concepts.

In this book, you will find a Mathematics Glossary and a Glossary of Illustrations that will provide you with some important mathematical definitions and concepts. There may also be other terms and concepts that you should review. By studying words and phrases such as "vertex," "coordinates," "inverse property," or "function," you will find it easier to express yourself in mathematical terms. For example, regarding the term "associative property," you might note:

Associative Property: The grouping of numbers does not affect the sum or product of the numbers. An example would be: $(3 + 4) + 5 = 3 + (4 + 5)$ or $(6 \times 10) \times 3 = 6 \times (10 \times 3)$.

You will remember information better by:

> • Writing the information more than once
> • Using examples
> • Discussing your examples with a parent,
>   friend, or teacher

Don't Forget Those Formulas!

Few students like to memorize, but those who memorize certain facts and formulas have better success on the Mathematics FCAT. Some multiple-choice and gridded-response items will ask you to identify a correct formula for solving a problem.

There are certain facts and formulas that you are expected to have memorized. Your teacher, in preparation for the FCAT, will review most of these important formulas. Just as with familiarizing yourself with mathematical vocabulary words, you should memorize formulas and facts to help you do your best. Keep in mind that memorization works best when you use the information over and over again in your school work.

Examples of what you should know include the following:

> • Calculating the volume of a prism and cylinder
> • Converting units of measurement, such as
>   meters into centimeters
> • Simplifying expressions
> • Finding the equation of a line
> • Solving for a particular variable

Multiple-Choice Questions

Use "Codes" to Make Better Guesses

You might find it helpful to use "codes" to rate multiple-choice answer choices. Using your pencil in the test booklet, you can mark the following codes beside each multiple-choice response to see which is the best choice.

An example of a code used by a ninth-grade student is given below.

(+) Put a "plus sign" by an answer choice if you are not sure if it is correct, but you think it might be correct;

(?) Put a "question mark" by an answer choice if you are not sure if it is the correct answer, but you don't want to rule it out completely;

(–) Put a "minus sign" by an answer choice if you are sure it is the wrong answer. (Then choose from the other answers to make an educated guess.)

Remember, it is fine to write in your test booklet. The space in the booklet is yours to use to help you do better on the FCAT. You will not have points counted off for using this coding system or creating your own system to help you on multiple-choice questions.

Answer Every Question

It is very important to answer as many multiple-choice questions as possible, even if you make an educated guess. On multiple-choice questions, you have a one in four chance of getting a question right, even if you just close your eyes and guess! This means that for every four questions you guess, the odds are you will get about one (25%) of the answers right. Guessing alone is not going to make you a star on the FCAT, but leaving multiple-choice questions blank is not going to help you either.

Take Advantage of "Chance"

On the Mathematics section of the FCAT, it is very important to answer as many multiple-choice items as possible, even if you make a well thought-out guess, because luck is with you! If you can eliminate even one possible answer, your chances of success are now even better! The best way to improve your chances on multiple-choice items is to use strategies such as using codes and power guessing that are described in this chapter. Learning how to improve your chances by using educated guessing is not cheating! In fact, you probably use this strategy outside of the classroom and don't even think about it. Imagine you have misplaced your favorite CD, and you want to find it before you leave for your friend's house. There are many possible places that it could be, but you use your common sense to eliminate some possibilities, thereby saving time searching and increasing your chances of finding it in time. For example, it might be possible that you left it in your sister's room, but you remember, "That isn't likely because her CD player has been broken for a month." That leaves you one less place to look, and more chances for success.

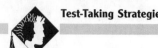

Understand Multiple-Choice Questions and Answer Choices

Each multiple-choice item will have four possible answers that follow either a question or a statement. For example, you might be asked, "The new school auditorium is 400 feet long and 200 feet wide, and the stage makes up 25% of the space. How much room is there for a set to be built?" You will be given four numbers to choose from to answer the question. Similarly, you could also be asked, "The band room is 2,500 square feet, and the stage is 25% of the space in the auditorium. Describe the size of the band room compared to the stage." Possible answers could be "double," "triple," "smaller," or "equal." Not all questions will ask for an exact number but all have just one correct answer. Even if the answers do not specify an exact number, it is assumed that you will have used mathematical reasoning to figure out a correct answer. Questions that do not ask for exact numbers are not easier questions.

Answer choices are not designed to be tricky, but they won't be easy to choose correctly by guessing. Answer choices will not have one answer that is obviously incorrect. For example, if you are asked about the average cost of five items with prices all under $20.00, you won't find a choice such as "$422."

Answer choices will contain answers that might seem correct if you made a common mistake in reasoning or calculation. For example, if you are asked the median age of teachers at your school, that calls for a different mathematical calculation than if you were asked for the mode age. There may be an answer choice that correctly reflects the mode age of teachers, but this would not be the correct answer. Some answer choices also reflect common miscalculations, such as adding when a number should be multiplied. When choosing answers, make sure to:

> - Carefully read the question
> - Check for any common or careless errors you may have made
> - Recheck your thinking and your calculations

Talk to Yourself!

Believe it or not, talking to yourself is a great FCAT Mathematics strategy! You may not think about it, but you probably talk to yourself all the time when you are solving problems in everyday life, especially problems that have "steps." Much of mathematics calls for "linear thinking," meaning that problems are best solved when certain steps are followed in order.

Imagine that you are having a problem with your computer. You are typing an essay that is due the next day, when suddenly your mouse will not work. If you are a successful problem solver you will first identify the problem: "The mouse isn't working." You then go through a checklist of how to proceed. "Let me see if there is a program running that is messing up my word processing." You hit a few keys and don't see any interference. The next step is to look at the wireless mouse. You say to yourself, "Let me see if there is a problem with the mouse." You pick it up and it looks fine. Then you say, "Well, something on my desk could be the problem." And you are right! Your big metal stapler is blocking the signal between your mouse and the computer. You solved the problem, step by step, by talking to yourself.

The same strategy works for mathematical problems. Suppose you were asked, "Do you want to buy a used car? You'll need a 20% down payment on the car. The car costs $3,000.00. How many dollars will you have to save to make the down payment?" Identify the question asked, "This question has something to do with percentages." Then work through the processes involved, "I need to find a formula to help me find a part of a whole." Then say to yourself, "I will translate this into a sentence such as, 'Twenty-percent is what part of $3,000.00?' " From that point on, there are a number of ways to solve this problem, and you can choose whatever is easier for you. For example, you could convert 20% to its decimal form of 0.20 and then multiply by $3,000.00. You can then evaluate your answer by verbally reviewing your reasoning and checking your work. So, if your answer is $6,000.00, instead of the correct answer of $600.00, you should be able to find your error by repeating your calculations to yourself and asking yourself if the answer makes sense.

Learn How to "Power Guess"

Not everything you know was learned in a classroom. Part of what you know comes from just living your day-to-day life. When you take the Mathematics section of the FCAT, you should use everything you have learned in school, but you should also use your experiences outside of the classroom to help you answer multiple-choice questions correctly. You might think to yourself, "Well, mathematics is different than other types of problems. There is just one correct answer. Things in mathematics are either right or wrong. If I don't know the answer right away, nothing else will help me get the question right." Although you might think that mathematics is different or harder, you still can use common sense thinking to help you do your best. Power guessing does not take the place of practice and knowledge, but it can help you to make reasonable choices by using what you know. Even if you eliminate one incorrect multiple-choice response, you have increased your chances of guessing correctly.

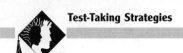

For example, take a look at this multiple-choice Mathematics question:

 Jeffrey's favorite music is "rock oldies," but his mother insists that the car radio be tuned to her favorite station, which plays "rock oldies" only about 12 minutes of every hour. What is the probability that Jeffrey's choice of music will be playing when his mom starts up the car?

    A. $p = 0.12$
    B. $p = 0.20$
    C. $p = 0.90$
    D. $p = 0.40$

Analysis: *Suppose you forgot the formula to find probability. However, you know that probability can be simply described as "the chance that something will happen." The question tells you that Jeffrey's favorite music isn't played very often on his mother's chosen radio station. Now take a look at the answers. You will see that answer "C" shows a large number, 0.90, which suggests that Jeffrey would have a good chance of hearing his favorite tunes. That does not make logical sense. You know that 0.90 is probably incorrect, just using what you know about chance. (For example, if the news reported a 90% chance of rain, you wouldn't choose to schedule a softball practice). Thus, you eliminate choice "C". You are now left with three choices, and your chances of guessing the correct answer are improved. And now that you are thinking about the answer, the formula might come to mind. (The correct answer is "B," which is found by converting an hour to 60 minutes and then dividing, $12 \div 60 = 0.20$.)*

Gridded-Response Questions

How to Do Great with Those Grids

On the FCAT you will be asked to show some of your answers using grids. A grid gives you an organized way to show what you know. Once you take the time to learn about grids you will find them easy to use and helpful for test success. Using grids can seem tricky at first but no one is trying to trick you. All it takes is a little practice and some special hints and you can be the "gridmaster" of the universe!

Writing First, Bubbles Second

After you have figured out a numerical answer and checked your work (don't forget to always check your work), it is time to write your answer in the spaces at the top of the grid, called answer boxes. Fill in the written answer exactly as you will fill in the grid. For example, if the answer is "679," do not write the number using just any answer boxes. Remember, you will then fill in the answer bubble directly below your written answer, so if you are sloppy in writing your answer, it will be almost certain that you will fill in the bubbles incorrectly.

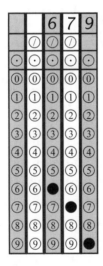
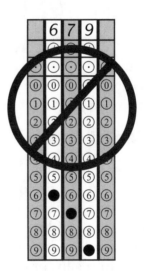

You will also see that some of the answer boxes and columns are shaded. This will help you easily see if the numbers in your answer are in the correct columns, or if you left an incorrect space blank. However, if you do not write your answers on the shaded areas dark enough to be read, you may end up filling in the wrong bubbles. Also be sure to fill in the bubbles with SOLID black marks that completely fill in the circle. Little dots, check marks, or incompletely filled bubbles will only lead to "bubble trouble"!

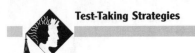
Right or Left, Left or Right, Either Way It Will Be All Right!

No matter what type of question, your answer will always fit into a five-column grid. You can choose to place the first digit of your answer in the left answer box column, or the last digit in the right answer box column. Some students feel more comfortable with one way or the other.

For example, let's say that the correct answer is "263." You could place the number "2" in the first column on the left side, or the "3" in the farthest right column. Both answers would be correct and would look like the grids below:

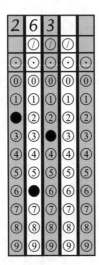 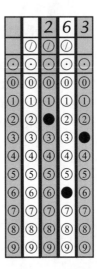

Be careful not to leave a blank answer box or column in the middle of an answer, although not all answers will use every space available.

Practice Makes Perfect

Many students think that filling in grids is no big deal but that type of thinking is not a good test-taking strategy. Once you begin to use grids, you will see that practice makes you more confident to show what you know by not wasting valuable test-taking time trying to remember how to fill in those grids. It can be frustrating to know the correct answer, but not know how to easily fill in the grids. So start with some simple problems and get comfortable filling in grids. It may sound silly, or even a little boring, but if filling in grids comes naturally, you can focus on finding the right answers instead of wondering if your answer will be scored correctly. Astronauts spend days learning how to use simple tools on a spacewalk, allowing them to concentrate on solving important problems. You may not be in space but your FCAT scores will soar if you practice filling in those grids!

Units? What Units?

Imagine that you are at home one day, the phone rings, and the caller says, "You are the grand prize winner of 100!" At first this sounds great! But then, wait…"A hundred of WHAT?" you ask yourself. A hundred dollars would be nice. A hundred feet of noodles may be less exciting. Paying attention to the units required in a gridded-response will help you on your way to test success. If the question asks for feet, then your answer needs to be in feet, not inches. While both answers may be correct, only the answer in feet will be scored as the correct answer.

What Are Those Slashes and Dots?

At first glance, the top part of a grid could look like some weird game of pool or checkers. Right under the gridded-response boxes are three bubbles with "slashes" and five bubbles with "dots." The bubbles with slashes are used for showing fractions, and the bubbles with dots are used to show decimals.

Decimals are shown on grids just like you would show them on paper. But remember that decimals take up spaces just as numbers would. If your answer starts with a decimal, and you choose to put the first number of your answer in the farthest space to the left, that space will be a decimal (see the example farhest to the left). If you choose to put the last number of your answer in the farthest right space, the decimal may or may not be in the last space on the left, depending on the length of your answer and the appropriate placement of the decimal in your answer.

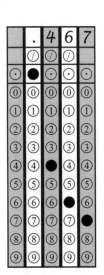
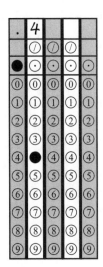

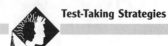
Fractions aren't hard to show either, provided that you do not write mixed numbers, such as $16\frac{1}{2}$, because this type of answer, although mathematically correct, cannot be scored using a grid. You must convert your answer to an improper fraction, such as $\frac{33}{2}$, or to a decimal such as 16.50. (This is a good reason to review the conversion of fractions.) The example below illustrates a correct answer using a mixed number.

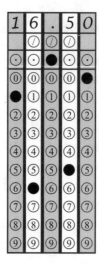

If asked, you can also express an answer in a percent. However, do not use decimals to show a percent. If you are trying to show 60%, or 60 percent, you could place your answer on the grid in these two ways:

When answering a gridded-response item remember to:

- Check your calculations

- Decide how to place your answer in the grid

- Carefully fill in the bubbles

- Make sure you use the correct units

- Double check to see that your answer is properly expressed in the grid

BLANK PAGE

Mathematics

Introduction

In the Mathematics section of the Florida Comprehensive Assessment Test (FCAT), you will be asked questions designed to assess the knowledge you have gained throughout your academic career. These questions have been constructed based on the mathematics skills you have been taught in school through the ninth grade. Within this section of the FCAT, your knowledge will be assessed with multiple-choice and gridded-response items. The questions are not meant to confuse you or to trick you, but are written so you have the best opportunity to show what you know about mathematics.

This *Show What You Know® on the FCAT Mathematics for Grade 9, Student Self-Study Workbook* contains the following:

• The Sunshine State Standards (Strands, Standard, and Benchmarks)

• The Mathematics Practice Tutorial with a sample question for each standard tested, sample responses, and an answer key with in-depth analyses.

• Two full-length Mathematics Assessments with sample responses, correlation charts, and answer keys with in-depth analyses.

Show What You Know® on the FCAT Mathematics for Grade 9, Student Self-Study Workbook will help you practice your test-taking skills. The Mathematics Practice Tutorial and two full-length Mathematics Assessments have been created to model the Grade 9 Florida Comprehensive Assessment Test in Mathematics.

Understanding Grade Level Expectations

Subject Area: MA: Mathematics

A **strand** is a category of knowledge. The five strands assessed on the FCAT Mathematics test are Number Sense, Concepts, and Operations, Measurement, Geometry and Spatial Sense, Algebraic Thinking, and Data Analysis and Probability.

Each Mathematics **standard** is a general statement of expected student achievement within a strand. The standards are the same for all grade levels.

Benchmarks are specific statements of expected student achievement under each Mathematics standard. Test items are written to assess the benchmarks. In some cases, two or more related benchmarks are grouped together because the assessment of one benchmark necessarily addresses another benchmark.

> **MA: Mathematics**
>
> **Strand A: Number Sense, Concepts, and Operations**
>
> **Standard 1:**
> **The student understands the different ways numbers are represented and used in the real world.**
>
> 1. Associates verbal names, written word names, and standard numerals with integers, rational numbers, irrational numbers, real numbers, and complex numbers.

The Grade Level Expectation's **Numbering System** identifies the Subject Area, the Strand, the Level, and the Benchmark. For example, in the Grade Level Expectation (MA.A.1.4.1), the first letters stand (MA) for the Subject Area, Mathematics, and the second letter (A.) for the Strand, (Number Sense). The first number (1.) stands for the Standard, the second number (4.) for the Level (Grades 9–12), the third number (1.) for the Benchmark. **Note:** The Grade Level Expectations are not intended to take the place of a curriculum guide, but rather to serve as the basis for curriculum development to ensure that the curriculum is rich in content and is delivered through effective instructional activities. The Grade Level Expectations are in no way intended to limit learning, but rather to ensure that all students across the state receive a good educational foundation that will prepare them for a productive life.

Sunshine State Standards

Mathematics

Strand A: Number Sense, Concepts, and Operations
   Standard 1: The student understands the different ways numbers are represented and used in the real world.

   MA.A.1.4.1 Associates verbal names, written word names, and standard numerals with integers, rational numbers, irrational numbers, real numbers, and complex numbers.

   MA.A.1.4.2 Understands the relative size of integers, rational numbers, irrational numbers, and real numbers.

   MA.A.1.4.3 Understands concrete and symbolic representations of real and complex numbers in real-world situations.

   MA.A.1.4.4 Understands that numbers can be represented in a variety of equivalent forms, including integers, fractions, decimals, percents, scientific notation, exponents, radicals, absolute value, and logarithms.

   Standard 2: The student understands number systems.

   MA.A.2.4.2 Understands and uses the real number system.

   Standard 3: The student understands the effects of operations on numbers and the relationships among these operations, selects appropriate operations, and computes for problem solving.

   MA.A.3.4.1 Understands and explains the effects of addition, subtraction, multiplication, and division on real numbers, including square roots, exponents, and appropriate inverse relationships.

   MA.A.3.4.2 Selects and justifies alternative strategies, such as using properties of numbers, including inverse, identity, distributive, associative, and transitive, that allow operational shortcuts for computational procedures in real-world or mathematical problems.

   MA.A.3.4.3 Adds, subtracts, multiplies, and divides real numbers, including square roots and exponents, using appropriate methods of computing, such as mental mathematics, paper and pencil, and calculator.

   Standard 4: The student uses estimation in problem solving and computation.

   MA.A.4.4.1 Uses estimation strategies in complex situations to predict results and to check the reasonableness of results.

Strand B: Measurement
Standard 1: The student measures quantities in the real world and uses the measures to solve problems.

MA.B.1.4.1 Uses concrete and graphic models to derive formulas for finding perimeter, area, surface area, circumference, and volume of two- and three-dimensional shapes, including rectangular solids, cylinders, cones, and pyramids.

MA.B.1.4.2 Uses concrete and graphic models to derive formulas for finding rate, distance, time, angle measures, and arc lengths.

MA.B.1.4.3 Relates the concepts of measurement to similarity and proportionality in real-world situations.

Standard 2: The student compares, contrasts, and converts within systems of measurement (both standard/nonstandard and metric/customary).

MA.B.2.4.1 Selects and uses direct (measured) or indirect (not measured) methods of measurement as appropriate.

MA.B.2.4.2 Solves real-world problems involving rated measures (miles per hour, feet per second).

Standard 3: The student estimates measurements in real-world problem situations.

MA.B.3.4.1 Solves real-world and mathematical problems involving estimates of measurements, including length, time, weight/mass, temperature, money, perimeter, area, and volume, and estimates the effects of measurement errors on calculations.

Strand C: Geometry and Spatial Sense
Standard 1: The student describes, draws, identifies, and analyzes two- and three-dimensional shapes.

MA.C.1.4.1 Uses properties and relationships of geometric shapes to construct formal and informal proofs.

Standard 2: The student visualizes and illustrates ways in which shapes can be combined, subdivided, and changed.

MA.C.2.4.1 Understands geometric concepts such as perpendicularity, parallelism, tangency, congruency, similarity, reflections, symmetry, and transformations including flips, slides, turns, enlargements, rotations, and fractals.

Standard 3: The student uses coordinate geometry to locate objects in both two and three dimensions and to describe objects algebraically.

MA.C.3.4.1 Represents and applies geometric properties and relationships to solve real-world and mathematical problems including ratio, proportion, and properties of right triangle trigonometry.

MA.C.3.4.2 Using a rectangular coordinate system (graph), applies and algebraically verifies properties of two- and three-dimensional figures, including distance, midpoint, slope, parallelism, and perpendicularity.

Strand D: Algebraic Thinking

Standard 1: The student describes, analyzes, and generalizes a wide variety of patterns, relations, and functions.

MA.D.1.4.1 Describes, analyzes, and generalizes relationships, patterns, and functions, using words, symbols, variables, tables, and graphs.

MA.D.1.4.2 Determines the impact when changing parameters of given functions.

Standard 2: The student uses expressions, equations, inequalities, graphs, and formulas to represent and interpret situations.

MA.D.2.4.1 Represents real-world problem situations using finite graphs, matrices, sequences, series, and recursive relations.

MA.D.2.4.2 Uses systems of equations and inequalities to solve real-world problems graphically, algebraically, and with matrices.

Strand E: Data Analysis and Probability

Standard 1: The student understands and uses the tools of data analysis for managing information.

MA.E.1.4.1 Interprets data that have been collected, organized, and displayed in charts, tables, and plots.

MA.E.1.4.2 Calculates measures of central tendency (mean, median, and mode) and dispersion (range, standard deviation, and variance) for complex sets of data and determines the most meaningful measure to describe the data.

MA.E.1.4.3 Analyzes real-world data and makes predictions of larger populations by applying formulas to calculate measures of central tendency and dispersion using the sample population data, and using appropriate technology, including calculators and computers.

Standard 2: The student identifies patterns and makes predictions from an orderly display of data using concepts of probability and statistics.

MA.E.2.4.1 Determines probabilities using counting procedures, table, tree diagrams, and formulas for permutations and combinations.

MA.E.2.4.2 Determines the probability for simple and compound events as well as independent and dependent events.

Standard 3: The student uses statistical methods to make inferences and valid arguments about real-world situations.

MA.E.3.4.1 Designs and performs real-world statistical experiments that involve more than one variable, then analyzes results and reports findings.

MA.E.3.4.2 Explains the limitations of using statistical techniques and data in making inferences and valid arguments.

About the FCAT Mathematics for Grade 9

Items in this section of the FCAT will test the student's ability to perform mathematical tasks in real-world and mathematical situations and will neither require students to define mathematical terminology nor to memorize specific facts. Ninth grade students are also allowed to use the Mathematics Reference Sheet during testing. The FCAT is meant to gauge a student's ability to apply mathematical concepts to a given situation.

The FCAT Mathematics does not contain any open-ended questions but contains multiple-choice questions, which require students to identify the correct answer out of four possible choices, and gridded-response questions, which require students to fill their responses into grids.

For multiple-choice questions, each question has only one correct answer; the other three choices are distractors representing incorrect answers that students commonly obtain for the question. Multiple-choice items are worth one point each. Students should spend no more than one minute answering each individual question, but they should be sure to allow themselves time to scrutinize each possible choice.

Gridded-response questions also have only one correct answer, but in certain circumstances, the answer may be represented in different formats. For example, if a question asks what fraction is equal to 50%, students may respond with 1/2, 4/8, 50/100, or any other fraction equaling 50%. Gridded-response questions are worth one point each.

Using the Skills Charts

The Skills Charts on pages 161–162 and 211–212 show the correct answers for each Assessment Test question, the standards covered by each question, and some keywords that indicate the subject covered by each question. Students can use the chart as they score their tests to see which standards they may need to review before taking the FCAT Mathematics Test.

Using the Correlation Charts

The Correlation Charts on pages 172–174 and 226–228 can be used by the teachers to identify areas of improvement. When students miss a question, place an "X" in the corresponding box. A column with a large number of "Xs" shows more practice is needed with that particular standard. Permission is granted by the publisher to reproduce the Correlation Charts to one teacher for use in a single classroom.

BLANK PAGE

Grade 9 FCAT Mathematics Reference Sheet

Area

Triangle $A = \frac{1}{2}bh$

Rectangle $A = lw$

Trapezoid $A = \frac{1}{2}h(b_1 + b_2)$

Parallelogram $A = bh$

Circle $A = \pi r^2$

KEY			
b	= base	d	= diameter
h	= height	r	= radius
l	= length	A	= area
w	= width	C	= circumference
ℓ	= slant height	V	= volume
$S.A.$	= surface area		

Use 3.14 or $\frac{22}{7}$ for π

Circumference

$C = \pi d$ or $C = 2\pi r$

Volume/Capacity   Total Surface Area

Right Circular Cone $V = \frac{1}{3}\pi r^2 h$ $S.A. = \frac{1}{2}(2\pi r)\ell + \pi r^2$ or $S.A. = \pi r\ell + \pi r^2$

Right Square Pyramid $V = \frac{1}{3}lwh$ $S.A. = 4(\frac{1}{2}l\ell) + l^2$ or $S.A. = 2l\ell + l^2$

Sphere $V = \frac{4}{3}\pi r^3$ $S.A. = 4\pi r^2$

Right Circular Cylinder $V = \pi r^2 h$ $S.A. = 2\pi rh + 2\pi r^2$

Rectangular Prism $V = lwh$ $S.A. = 2(lw) + 2(hw) + 2(lh)$

In the following formulas, n represents the number of sides.
- In a polygon, the sum of the measures of the interior angles is equal to $180(n - 2)$.
- In a regular polygon, the measure of an interior angle is equal to $\frac{180(n - 2)}{n}$.

Grade 9 FCAT Mathematics Reference Sheet

Pythagorean theorem:

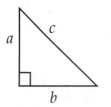

$$a^2 + b^2 = c^2$$

Slope-intercept form of an equation
of a line:

$$y = mx + b$$

where m = slope and b = the y-intercept.

Distance, rate, time formula:

$$d = rt$$

where d = distance, r = rate, t = time.

Distance between two points
$P_1 (x_1, y_1)$ and $P_2 (x_2, y_2)$:

$$\sqrt{(x_1 - x_2)^2 + (y_1 - y_2)^2}$$

Midpoint between two points
$P_1 (x_1, y_1)$ and $P_2 (x_2, y_2)$:

$$\left(\frac{x_2 + x_1}{2} , \frac{y_2 + y_1}{2} \right)$$

Simple interest formula:

$$I = prt$$

where p = principal, r = rate,
t = time.

Fold and Tear Carefully Along Dotted Line.

Conversions

1 yard = 3 feet = 36 inches
1 mile = 1760 yards = 5280 feet
1 acre = 43,560 square feet
1 hour = 60 minutes
1 minute = 60 seconds

1 cup = 8 fluid ounces
1 pint = 2 cups
1 quart = 2 pints
1 gallon = 4 quarts

1 liter = 1000 milliliters = 1000 cubic centimeters
1 meter = 100 centimeters = 1000 millimeters
1 kilometer = 1000 meters
1 gram = 1000 milligrams
1 kilogram = 1000 grams

1 pound = 16 ounces
1 ton = 2000 pounds

Metric numbers with four digits are presented without a comma (e.g., 9960 kilometers).
For metric numbers greater than four digits, a space is used instead of a comma
(e.g., 12 500 liters).

Examples of Common Two-Dimensional Shapes

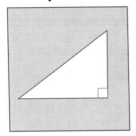

Right Triangle

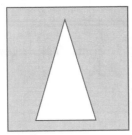

Isosceles Triangle

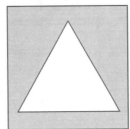

Equilateral Triangle

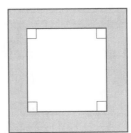

Square

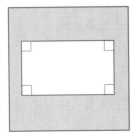

Rectangle

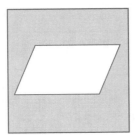

Parallelogram

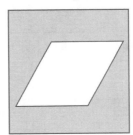

Rhombus

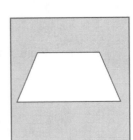

Trapezoid

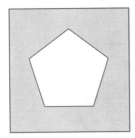

Pentagon

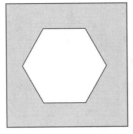

Hexagon

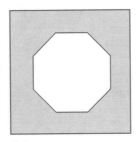

Octagon

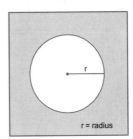

Circle

Examples of Common Three-Dimensional Shapes

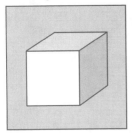

Cube

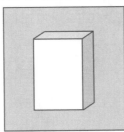

Rectangular Prism

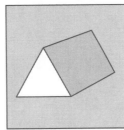

Triangular Prism

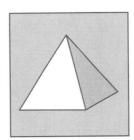

Pyramid

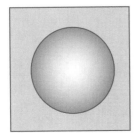

Sphere

Cylinder

Cone

Examples of How Lines Interact

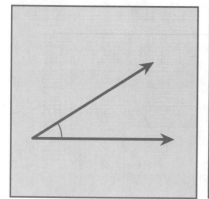

Acute Angle

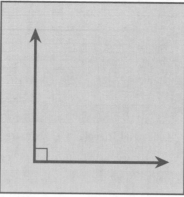

Right Angle

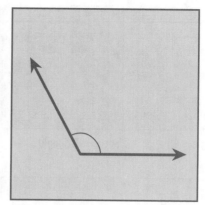

Obtuse Angle

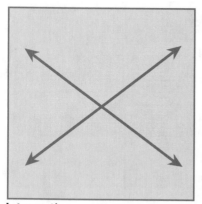

Intersecting

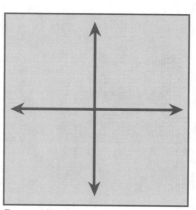

Perpendicular

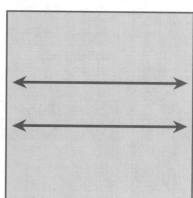

Parallel

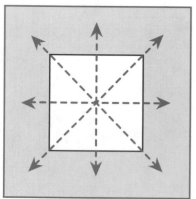

Lines of Symmetry

Examples of Types of Graphs

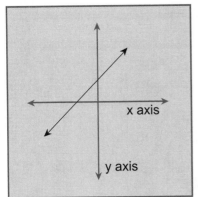

Line Graph

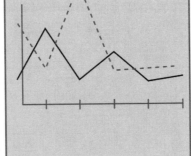

Double Line Graph

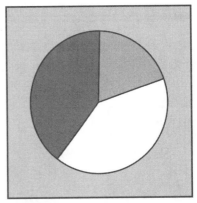

Pie Chart

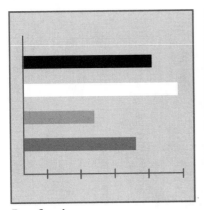

Bar Graph

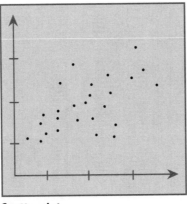

Scatterplot

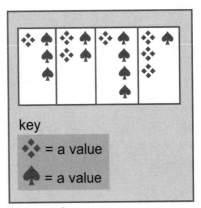

key

◆ = a value

♠ = a value

Pictograph

3	3 4
4	8
5	1 9
6	2 6 6
7	0 1 5

Stem and Leaf Plot

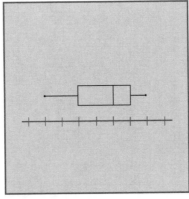

Box and Whisker

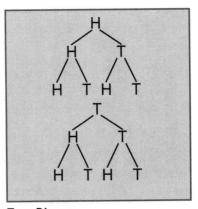

Tree Diagram

Examples of Object Movement

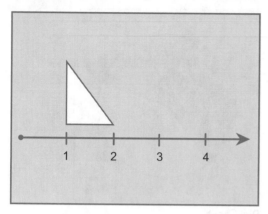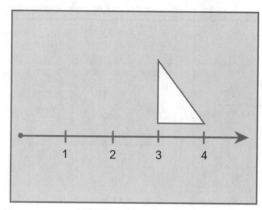

Translation

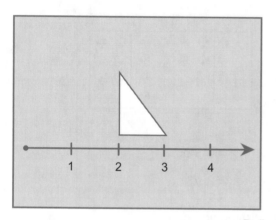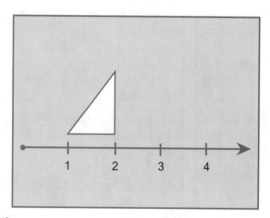

Reflection

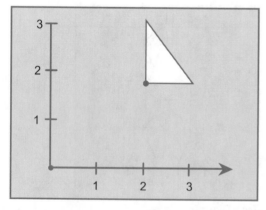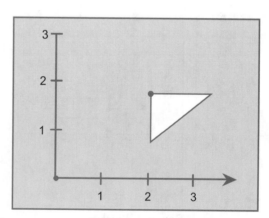

Rotation

GLOSSARY

absolute value: The non-negative value of any number; the value of any number disregarding its sign; or the distance a number is from zero on a number line; denoted as |n| for any real number n. For example, the absolute value of 4 is 4 and the absolute value of -4 is also 4.

acute angle: An angle that measures less than 90 degrees and greater than 0 degrees.

acute triangle: A triangle with three acute angles.

addition: An operation joining two or more sets where the result is the whole.

altitude: An line drawn from any vertex of a polygon to any side so that the line is perpendicular to the side to which it is drawn. In a three-dimensional figure, it is a line drawn from any vertex of the solid to any face so that the line is perpendicular to the face to which it is drawn. Also known as the height.

analyze: To break down material into component parts so that it may be more easily understood.

angle: The distance, recorded in degrees (°), between two segments, rays, or lines that meet at a common vertex. Angles can be obtuse, acute, right, or straight.

approximate: To obtain a number close to an exact amount.

approximation: The result of obtaining a number close to an exact amount.

area: The amount of two-dimensional space enclosed by a flat object is referred to as its area. The units used to measure area are always some form of square units, such as square inches or square meters. The most common abbreviation for area is A.

argument: A reason or reasons offered for or against something; suggests the use of logic and facts to support or refute a statement or idea.

associative property: This property states that the addition or multiplication of three or more numbers will result in the same sum (in addition) or product (in multiplication) regardless of how the numbers are grouped. For example, any numbers a, b, and c, in addition: (a + b) + c = a + (b + c); for multiplication: (a x b) x c = a x (b x c).

attribute: A characteristic or distinctive feature.

average: A measure of central tendency; generally, the word average implies the mean or arithmetic average, but it could also refer to the median or mode.

average: The sum of a set of numbers divided by the total number of terms in the set. For example, the average of the numbers 1, 2, and 6 is (1 + 2 + 6) / 3, which equals 3. *See mean.*

axes: Perpendicular lines used as reference lines in a coordinate system or graph; traditionally, the horizontal axis (*x*-axis) represents the independent variable and the vertical axis (*y*-axis) the dependent variable.

bar graph: A graph that uses the lengths of rectangular bars to represent numbers and compare data.

base: Usually refers to the side of a polygon closest to the bottom of the page. In a triangle, the other two sides are called legs. Also, the face around which a three-dimensional object is formed. For example, the base of a triangular prism is a triangle, and the base of a square pyramid is a square.

GLOSSARY

box-and-whisker plot: A type of graph depicted on a number line that is used to express statistical data. The box portion of the graph represents the middle 50% of the given values. The vertical line within the box denotes where the median of the data set falls. The whiskers are extended outward in both directions from the graph and represent the upper and lower 25% of the given values. The extreme points of the whiskers represent the minimum and maximum values of the given data.

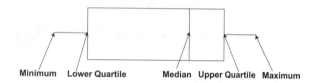

Minimum   Lower Quartile   Median   Upper Quartile   Maximum

chart: A method of displaying information in the form of a graph or table.

circle: A set of points in a plane that are all the same distance from the center point.
Example: A circle with center point P is shown below.

circle graph: Sometimes called a pie chart; a way of representing data that shows the fractional part or percentage of an overall set as an appropriately-sized wedge of a circle. Example:

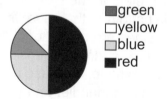

☐ green
☐ yellow
☐ blue
■ red

circumference: The boundary line, or perimeter, of a circle; also, the length of the perimeter of a circle. Example:

common denominator: A number divisible by all of the denominators of two or more fractions. Example: The fractions 1/2, 1/3, 1/4 have a common denominator of 12 because 12 is divisible by 2, 3, and 4.

commutative property: This property states that the addition or multiplication of two numbers will result in the same sum (in addition) or product (in multiplication) regardless of the order of the two numbers. For example, any numbers a and b, in addition: a + b = b + a; in multiplication: a x b = b x a.

compare: Look for similarities and differences.

complementary angles: Two angles with a sum of 90°.

composite number: A number that has more than two factors is called a composite number. Examples include 4, 35, and 121. The numbers zero and one are not composite numbers. *See prime number.*

conclude: To make a judgment or decision after investigating or reasoning; to infer.

conclusion: A statement that follows logically from other facts.

GLOSSARY

cone: A three-dimensional figure with one circular or elliptical base and a curved surface that joins the base to a single point called the vertex.

cones

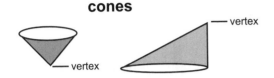

congruent figures: Figures that have the same shape and size.

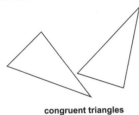

congruent triangles

coordinates: Ordered pairs of numbers that identify the location of points on a coordinate plane. Example: (3, 4) is the coordinate of point A.

contrast: Look for differences.

cross-multiply: A method used to solve proportions or evaluate whether or not two fractions are equal. In order to cross-multiply, you must have one fraction on each side of an equals sign to make an equation. Form two products by multiplying each numerator by the denominator of the other fraction: For example, if you are asked whether or not 3/4 equals 6/8, set them equal to one another and cross multiply:

$$\frac{3}{4} \diagdown\diagup \frac{6}{8}; \quad 3 \times 8 = 4 \times 6; \quad 24 = 24.$$

Since these products are equal, the fractions are also equal.

cube: A rectangular prism having six congruent square faces.

customary system: *See U.S. system of measurement.*

cylinder: A solid figure with two circular or elliptical bases that are congruent and parallel to each other connected by a curved lateral area.

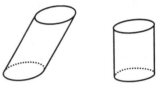

data: Collected pieces of information.

decimal number: A number expressed in base 10, such as 39.456 where each digit's value is determined by multiplying it by some power of ten

denominator: The number in a fraction below the bar; indicates the number of equivalent pieces or sets into which something is divided.

diagonal: A segment joining 2 non-consecutive vertices of a polygon.

diagram: A drawing that represents a mathematical situation.

diameter: A line segment (or the length of a segment) passing through the center of the circle with end points on the circle.

difference: The number found when subtracting one number from another; the result of a subtraction operation; the amount by which a quantity is more or less than another number.

dimensions: The length, width, or height of an object.

GLOSSARY

direct proportion: A method of comparing or solving two equal ratios. For example, if for every 3 people there are 2 dogs, how many dogs will there be for 9 people? The direct proportion to solve this sets the original/known ratio of people to dogs equal to the unknown ratio: $\frac{3}{2} \diagdown \frac{9}{n}$; or $\frac{3}{2} \diagdown \frac{9}{6}$;. In this example, the answer derived from the direct proportion is 6. *See cross-multiply.*

distributive property: This property states that the same final answer will be found when a number is multiplied by the sum of two numbers, or when the first number is multiplied by both of the numbers separately and then the products are added together. For example, any numbers a, b, and c: a x (b + c) = a x b + a x c.

dividend: A number which is to be divided by another number. Dividend ÷ divisor = quotient. Example: In 15 ÷ 3 = 5, 15 is the dividend.

$$\overset{\text{quotient}}{\text{divisor}) \overline{\text{dividend}}} \qquad 3\overline{)\overset{5}{15}}$$

divisible: One integer is divisible by another non-zero integer if the quotient is an integer with a remainder of zero. Example: 12 is divisible by 3 because 12 ÷ 3 is an integer, namely 4.

division: An operation on two numbers to determine the number of sets or the size of the sets. Problems where the number of sets is unknown may be called measurement or repeated subtraction problems. Problems where the size of sets is unknown may be called fair sharing or partition problems.

divisor: The number by which the dividend is to be divided; also a factor quotient. Example: In 15 ÷ 3 = 5, 3 is the divisor.

$$\overset{\text{quotient}}{\text{divisor}) \overline{\text{dividend}}} \qquad 3\overline{)\overset{5}{15}}$$

edge: The line segment formed by the intersection of two faces of a three-dimensional figure; a cube has 12 edges.

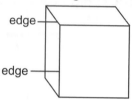

equality: Two or more sets of values that are equal.

equally likely: Two outcomes having the same probability of occurring.

equation: A number sentence or algebraic sentence which shows equality between two sets of values. An equation can be recognized by the presence of an equal sign (=). Examples: 4 + 8 = 6 + 6, 4 + 8 = 24 ÷ 2, 4 + x = 12

equiangular: In any given polygon, if the measures of all of the angles formed by the figure's segments are of equal value (congruent), the polygon is said to equiangular. All regular polygons are equiangular. Not all polygons that are equiangular are necessarily equilateral. This is illustrated by the diagram below.

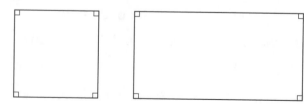

GLOSSARY

equilateral: Having equal sides. All regular polygons are equilateral. Not all polygons that are equilateral are necessarily equiangular.

estimate: To find an approximate value or measurement of something without exact calculation.

estimation: The process of finding an approximate value or measurement of something without exact calculation.

- Measurement estimation—an approximate measurement found without taking an exact measurement.
- Quantity estimation—an approximate number of items in a collection.
- Computational estimation—a number that is an approximation of a computation that we cannot (or do not wish to) determine exactly.

even number: A whole number divisible by two. Example: 0, 4, and 678 are even numbers.

event: Any subset of the sample space. In rolling a number cube, the event of rolling a "3" is a singleton event because it contains only one outcome. The event of rolling an "even number" contains three outcomes.

exponent: When a number is raised to a power, the power it is raised to is expressed by using an exponent. The exponent shows how many times a number is to be multiplied by itself. An exponent can be any rational number, but for the purpose of the FCAT, exponents will only include whole numbers greater than zero. When written, exponents appear after the number they influence and are slightly raised above the number, such as $2^3 = 2 \times 2 \times 2$ or $4^6 = 4 \times 4 \times 4 \times 4 \times 4 \times 4$, where 3 and 6 are the exponents.

expanded form: A number written in component parts showing the cumulative place values of each digit in the number. Example: 546 = 500 + 40 + 6.

expression: A combination of variables, numbers, and symbols that represent a mathematical relationship.

extrapolate: To make a guess about a value, function, or graph beyond the values already known. For example, given the sequence of numbers {. . . 2, 4, 6, 8, . . .}, you may predict based on the observed pattern, that the next number in the sequence after 8 is 10 and the number before 2 is 0.

face: A flat surface, or side, of a solid (3-D) figure. This square pyramid has four triangular faces and one square face also called its base.

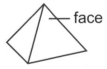
face

factor: One of two or more numbers that are multiplied together to obtain a product, or an integer that you can divide evenly into another number. Example: In 4 x 3 = 12, 4 and 3 are factors of 12.

figure: A geometric figure is a set of points and/or lines in 2 or 3 dimensions.

flip: Movement of a figure or object over an imaginary line of symmetry that reverses it producing a mirror image. Also called a reflection. Examples: Flipping a pancake from one side to the other. Reversing a "b" to a "d". Tipping a "p" to a "b" or a "b" to a "p" as shown below:

fraction: A way of representing part of a whole set. Example:

$$\frac{\text{numerator}}{\text{denominator}} = \frac{\text{dividend}}{\text{divisor}} =$$

$$\frac{\text{\# of parts under consideration}}{\text{\# of parts in a set}}$$

function: A relation, such as a graph, in which a variable, called the dependent variable, is dependent on another value, usually an independent variable. In a function, each value of x corresponds to only one value of y.

function machine: Applies a function rule to a set of numbers, which determines a corresponding set of numbers.
Example: Input 9 —> Rule x 7 —> Output 63. If you apply the function rule "multiply by 7" to the values 5, 7, and 9, the corresponding values would be

$$5 \longrightarrow 35$$
$$7 \longrightarrow 49$$
$$9 \longrightarrow 63$$

graph: A "picture" showing how certain facts are related to each other or how they compare to one another. Some examples of types of graphs are line graphs, pie charts, bar graphs, scatterplots, and pictographs.

greatest common factor (divisor): The largest factor of two or more numbers; often abbreviated as GCF. The GCF is also called the greatest common divisor.
Example: To find the GCF of 24 and 36:
1) Factors of 24 = {1, 2, 3, 4, 6, 8, 12, 24}.
2) Factors of 36 = {1, 2, 3, 4, 6, 9, 12, 18, 36}.
3) Common factors of 24 and 36 are {1, 2, 3, 4, 6, 12}, the largest being 12.
4) 12 is the GCF of 24 and 36.

grid: A pattern of regularly spaced horizontal and vertical lines on a plane that can be used to locate points and graph equations.

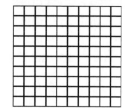

height: *See altitude.*

hexagon: A six-sided polygon. The total measure of the angles within a hexagon is 720°.

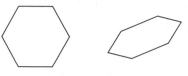

regular hexagon   nonregular hexagons

histogram: A graph that shows the frequency distribution for a set of data. A bar represents a range of values and there are no spaces between successive bars.

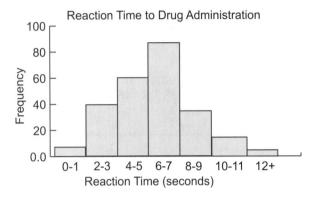

hypotenuse: In a right triangle, the side opposite the right angle. It is always the longest side of a right triangle.

identity property: This property states that, in addition, the sum of a number and zero will always equal the given number. In multiplication, the product of a given number and one will always equal the given number. For example, for any number a, in addition, $a + 0 = 0 + a = a$; in multiplication, $a \times 1 = 1 \times a = a$.

improper fraction: Any fraction in which the numerator has a higher absolute value than the denominator is called an improper fraction. All improper fractions can be converted to mixed numbers. Examples of improper fractions include 9/5, 26/11, 2/1, or 100/10.

inequality: Two or more sets of values are not equal. There are a number of specific inequality types, including less than (<), greater than (>), and not equal to (≠).

integer: Any number, positive or negative, that is a whole number distance away from zero on a number line, in addition to zero. Specifically, an integer is any number in the set {. . .-3,-2,-1, 0, 1, 2, 3. . .}. Examples of integers include 1, 5, 273, -2, -35, and -1,375.

intercept: A point on a graph where the line crosses the y-axis or x-axis. For a linear equation, the intercept occurs when one of the variables is equal to 0.

intersecting lines: Lines that meet at a point.

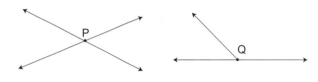

interval: Spacing of (or space between) two numbers on a number line.

inverse property: The property states that, in addition, the sum of a positive number and a negative number with the same absolute values will be zero. $4 + (-4) = 0$. In multiplication, the product of two reciprocal fractions will be one. $\frac{3}{4} \times \frac{4}{3} = 1$. For example, any number a, in addition, $a + -a = 0$. In multiplication, for any number $b \neq 0$. $b \times \frac{1}{b} = 1$.

irrational numbers: Any real number that cannot be expressed as the ratio of two integers is considered irrational. Some common examples include:

$$\sqrt{2}, \ \sqrt{3}, \text{ and } \pi.$$

isosceles triangle: A triangle with exactly two sides of equal length.

least common denominator: The smallest number divisible by all of the denominators of two or more fractions. Example: For 1/12, 3/4, and 2/3, 12 is the least common denominator because 12 is divisible by 12, 4, and 3.

GLOSSARY

justify: To prove or show to be true or valid using logic and/or evidence.

least common multiple (LCM): The smallest positive multiple of two or more integers. Example: The number 12 is the LCM of 3, 2, and 4, because it is the smallest number that is a multiple of all three numbers. 12 is also the LCM of 2, -3, 4.

line: As a working definition, think of it as one of the so-called undefined terms. A series of points that extend infinitely in two opposing directions.

line graph: A graph that uses lines, segments, or curves to show that something is increasing, decreasing, or staying the same over time. Note: A line graph does not have to be a straight line

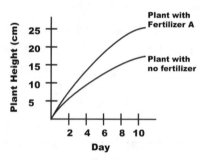

line of symmetry: A line on which a figure can be folded into two parts that are congruent mirror images of each other, so that every point on each half corresponds exactly to its image on the other half.

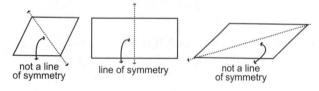

line plot: A line plot, sometimes called a dot plot, starts with a line that represents one variable. The values of the variable are labels on the line. Each observation is marked as a point above the line.

Line Plot for Quality Ratings for Natural Peanut Butter

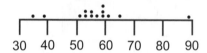

mean: A measure of central tendency found by adding the members of a set of data and dividing the sum by the number of members of the set (also called the arithmetic mean). Example: If A = 20 children, B = 29 children, and C = 26 children, the mean number of children is found by adding the three numbers (20 + 29 + 26 = 75) and then dividing the sum, 75, by the number 3. So, 25 is the mean of 20, 29, 26.

median: The number in the middle of a set of data arranged in order from least to greatest or from greatest to least; or the average of the two middle terms if there is an even number of terms. Example: For the data 6, 14, 23, 46, 69, 72, 94: the median is 46 (the middle number). For the data 6, 14, 23, 69, 72, 94: the median is also 46 (the average of the two middle numbers in the list).

method: A systematic way of accomplishing a task.

GLOSSARY

metric system: A measurement system based on the powers of ten. The following is a list of the base units of the metric system, as well as a few of their more common derivatives and abbreviations: length: millimeter, centimeter, meter, kilometer (mm, cm, m, km); volume: cubic centimeter, milliliter, liter (cc, ml, l); weight: grams, kilograms (g, kg); temperature: degrees Celsius (ºC); *see U.S. system of measurement.*

midpoint: A point on a line segment that divides the segment into two congruent parts.

mixed number: A number expressed as the sum of an integer and a proper fraction; having a whole part and a fractional part. Example: $6\frac{2}{3}$

mode: The item that occurs most frequently in a set of data. There may be one, more than one, or no mode.
Example: The mode in {1, 3, 4, 5, 5, 7, 9} is 5.

multiple: A multiple of a number is the product of that number and an integer.
Example: Multiples of 2 = {2, 4, 6, 8, 10, 12,....}. Multiples of 3 = {3, 6, 9, 12,....}. Multiples of 4 = {4, 8, 12,....}.

multiplication: An operation on two numbers that tells how many in all. The first number is the number of sets and the second number tells how many in each set. Problem formats can be expressed as repeated addition, an array, or a Cartesian product.

natural numbers: The set of integers used for counting, {1, 2, 3, 4, 5, . . .}.

number line: A line that shows numbers ordered by magnitude from left to right or bottom to top; equal intervals are marked and usually labeled.

number sentence: An expression of a relationship between quantities as an equation or an inequality. Examples: 7 + 7 = 8 + 6; 14 < 92; 56 + 4 > 59.

numerator: The number above the fraction bar in a fraction; indicates the number of equivalent parts being considered.

obtuse angle: An angle with a measure greater than 90 degrees and less than 180 degrees.

obtuse triangle: A triangle with one obtuse angle.

octagon: An eight-sided polygon. The total measure of the angles within an octagon is 1080º.

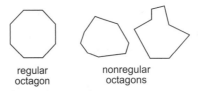

regular octagon   nonregular octagons

odd number: A whole number that is not divisible by two. Examples: The numbers 53 and 701 are odd numbers.

operation: A mathematical process that combines numbers; basic operations of arithmetic include addition, subtraction, multiplication, and division.

GLOSSARY

order of operations: In simplifying an expression involving a number of indicated operations, perform the operations in the following order:

1. Complete all operations inside parentheses first;
2. Calculate powers and roots in the order they occur from left to right;
3. Calculate all multiplications and divisions from left to right;
4. Calculate all additions and subtractions from left to right.

Examples: $7 + 3 \times 8 = 31$ [multiply 3×8 before adding 7]; $(7 + 3) \times 8 = 80$ [add 7 and 3 before multiplying by 8]; $7 + 3^2 \times 8 = 79$ [square 3, multiply by 8, and then add 7]. Sometimes noted with the acronymn PEMDAS.

ordered pairs: Two numbers (elements) for which order is important. When used to locate points on a coordinate graph, the first element indicates distance along the *x*-axis (horizontal), and the second indicates distance along the *y*-axis (vertical).

origin: Zero on a number line or the point (0, 0) on a coordinate plane.

parallel: Two lines, segments, or rays in the same plane that never intersect no matter how far they are extended. The symbol denoting parallel lines is ||.

parallelogram: A quadrilateral with opposite sides parallel.

pattern: An arrangement of numbers, pictures, etc., in an organized and predictable way. Examples: 3, 6, 9 12 or ® 0 ® 0 ® 0.

pentagon: A five-sided polygon. The total measure of the angles within a pentagon is 540º.

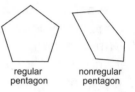

regular pentagon   nonregular pentagon

percent: A ratio of a number to 100. Percent means per hundred and is represented by the symbol %. Example: "35 to 100" means 35%.

perimeter: The total length of the outside border of an object is called its perimeter. For any polygon, the actual value is determined by finding the sum of the lengths of all of its sides. For example, a triangle with sides of 5 inches, 4 inches, and 3 inches has a perimeter of 12 inches. The units of measurement used to express perimeter are linear units, such as inches or kilometers. The most common abbreviation for perimeter is P.

perpendicular lines: Lines that lie on the same plane and intersect to form right angles (90 degrees).

90°
90°

pi: A number that expresses the ratio of the circumference of a circle to its diameter. It is represented by the symbol π and is used in circles to find both area and circumference. The exact value of pi cannot be expressed in our number system, but is approximately equal to 3.14 or 22/7.

GLOSSARY

pictograph: Graph that uses pictures or symbols to represent similar data. The value of each picture is interpreted by a "key" or "legend."

Key
Each picture =
10 pieces of fruit

pie chart: *See circle graph.*

place value: The value of a digit as determined by its place in a number. Example: In the number 135, the 3 means 3 x 10 or 30. In the number 356, the 3 means 3 x 100 or 300.

plane: One of the so-called undefined terms. As a working definition, think of it as any region that can be defined by a minimum of three noncollinear points and that extends infinitely in a two-dimensional manner.

plane figure: Any arrangement of points, lines, or curves within a single plane, a "flat" figure.

plot: To place points at their proper coordinates on a graph.

point: One of the so-called undefined terms. As a working definition, think of it as a location on a graph defined by its position in relation to the *x*-axis and *y*-axis. Points are sometimes called ordered pairs and are written in this form: (*x*-coordinate, *y*-coordinate).

polygon: A closed plane figure having three or more straight sides that meet only at their endpoints. Special polygons that have equal sides and equal angles are call regular polygons.

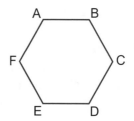

ABCDEF is a polygon.

population: A group of people, objects, or events that fit a particular description.

power: When a number is raised by an exponent, it is said to be raised by that power. For example, 3^4 can be thought of as the fourth power of three. To find the actual value, multiply the base number, in this case 3, by itself the number of times equal to the value of the exponent, in this case 4, which equates to 3 x 3 x 3 x 3. Therefore, the fourth power of three equals 81.

predict: To tell about or make known in advance, especially on the basis of special knowledge or inference.

prediction: A prediction is a description of what will happen before it happens. It is a foretelling that is based on a scientific law or mathematical model.

prime numbers: A whole number greater than 1 having exactly two whole number factors, itself and 1. Examples: The number 7 is prime since its only whole number factors are 1 and 7. One is not a prime number.

GLOSSARY

prism: A three-dimensional figure that has two congruent and parallel faces (bases) that are polygons; the remaining (lateral) faces are parallelograms. The volume of any right prism is found by multiplying the area of its base, *B*, by its height, *h*. (*V = Bh*)

probability: The numerical measure of the chance that a particular event will occur, depending on the possible events. The probability of an event, P(E), is always between 0 and 1, with 0 meaning that there is no chance of occurrence and 1 meaning a certainty of occurrence.

product: The result of a multiplication expression; factor x factor = product. Example: In 3 x 4 = 12, 12 is the product.

proper fraction: Any fraction with the numerator less than the denominator is called a proper fraction. By definition, the value of all proper fractions is less than one. Examples of proper fractions include 1/2, 5/16, 786/5563, and 22/144.

properties: Known interactions of numbers in specific situations. *See associative property, commutative property, distributive property, identity property, inverse property and zero property.*

proportion: *See direct proportion.*

pyramid: A solid (3-D) figure whose base is a polygon and whose other faces are triangles that meet at a common point called the vertex which is away from the base.

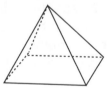

Pythagorean theorem: In any right triangle, the sum of the squares of the lengths of the two sides perpendicular to one another is equal to the square of the length of the hypotenuse: $a^2 + b^2 = c^2$.

quadrant: Any one of four unique sections of a two-dimensional graph. Quadrant I contains the points for which *x* and *y* are both positive; Quadrant II contains the points for which *x* is negative and *y* is positive; Quadrant III contains the points for which *x* and *y* are both negative; and Quadrant IV contains the points for which *x* is positive and *y* is negative.

quadrilateral: A four-sided polygon. Some types of quadrilaterals have special names and properties, including rectangles, squares, parallelograms, rhombi, and trapezoids. The total measure of the angles within a quadrilateral is 360°. Example: ABCD is a quadrilateral.

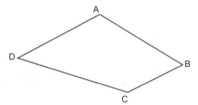

questionnaire: A set of questions for a survey.

quotient: The result of dividing one number by another number. Dividend ÷ divisor = quotient. Example: In 15 ÷ 3 = 5, 5 is the quotient.

radical: A mathematical operation symbolized by $\sqrt{}$. A radical is any number or expression that has a root. *See square root.*

radius: The distance from the center to the edge of a circle; or, the distance from the center of a circle to a point on the circle.

GLOSSARY

range: In a set of numbers, the difference between the two extremes in the set; in other words, the maximum value in a set minus the minimum value in a set. For example, the range of the set {2, 5, 8, 23, 46} is 46 – 2 = 44.

rate: A rate is an expression of how long it takes to do something. Examples of rates are miles per hour and revolutions per minute. In general, rate is measured as the number of times an event occurs divided by a unit of time.

ratio: A comparison of two numbers using a variety of written forms. Example: The ratio of two and five may be written "2 to 5" or 2:5 or 2/5.

rational number: A number that can be expressed as the ratio of two integers. Examples include 2 (written as a ratio: 2/1), .5 (written as a ratio: 1/2), and 1.75 (written as a ratio: 7/4).

ray: One of the so-called undefined terms. As a working definition, think of it as a half-line. A straight line extending infinitely in one direction from a given point.

real number: Any number that is either rational or irrational is in the set of real numbers. Real numbers are any values you might come across in real life. Examples of real numbers include -2, 0, 0.15, 1/2, $\sqrt{3}$, π, and 7.89 x 109.

reciprocal: When you take any value and raise it to the power of -1, you get its reciprocal. An easy way to think of it is to take the value and make it the denominator of a fraction with a numerator of 1. For example, the reciprocal of 3 is 1/3. Conversely, the reciprocal of 1/3 is 3.

rectangle: A quadrilateral with four right angles. A square is one example of a rectangle.

reduce: To put a fraction into its simplest form by dividing out any common factors. For example, 4/8 reduces to 1/2 in its simplest form.

reflection: A transformation of a figure created by flipping the figure over a line, creating a mirror image. *See flip.*

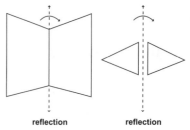

reflection   reflection

regular polygon: A special type of polygon that is both equilateral and equiangular.

represent: To present clearly; describe; show.

rhombus: A quadrilateral with all four sides equal in length. A square is a special type of rhombus

right angle: An angle whose measure is 90 degrees. The lines or segments which form right angles are said to be perpendicular to one another. *See angle and triangle.*

GLOSSARY

right triangle: A triangle having one right angle. S*ee angle and triangle.*

rotation: Moving an object around an imaginary point in a circular motion either clockwise or counterclockwise. After the move, the object will have the same shape and size but may be facing a different direction. *See turn.*

rounding: Taking an exact value and making it an approximation. Rounding is done by examining the value of the number in the place value to the right of the place value to which you want to round. If this number is less than 5 in value, you round down; if it is equal to 5 or greater, you round up.

rule: A procedure; a prescribed method; a way of describing the relationship between two sets of numbers. Example: In the following data, the rule is to add 3:

Input	Output
3	6
5	8
9	12

ruler: A straight-edged instrument used for measuring the lengths of objects. A ruler usually measures smaller units of length, such as inches or centimeters.

sample: A portion of a population or set used in statistics. Example: All boys under the age of ten constitute a sample of the population of all male children.

sample space: A set of all possible outcomes of a specified experiment.

scale: Sequenced collinear marks, usually at regular intervals or else representing equal steps, that are used as a reference in making measurements.

scatterplot: A type of graph containing points in which coordinates represent paired values. This type of graph is usually used when data seems more random than ordered.

scientific notation: A method of writing any rational number as a decimal number multiplied by some power of ten. It is most often used to represent very large or very small numbers. example: $138,000,000,000 = 1.38 \times 10^{11}$.

segment: One of the so-called undefined terms. As a working definition, think of it as a part of a line ending at specific points. Segments meet at vertices to form closed figures, that are both two- and three-dimensional.

semicircle: Half of a circle with the diameter as its base.

sequence: A set of numbers arranged in a special order or pattern.

set: Any grouping of numbers. A set can be specific or random, small or large. Sets are usually notated by placing numbers within brackets, as with {1, 2, 3}. Before finding statistical data of sets, you should always arrange the values in descending or ascending order.

side: A line segment connected to other segments to form the boundary of a polygon.

GLOSSARY

similar: Similar polygons must have the same shape, but not necessarily the same size.

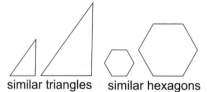

similar triangles　　similar hexagons

In order for two polygons to be similar, they must meet two conditions: 1. all pairs of corresponding angles must be congruent and 2. all pairs of corresponding sides must be proportional. This means that the ratio of the lengths of a pair of corresponding sides must be the same for all other pairs of corresponding sides.

simplify: To take a given mathematical expression and put it into its most basic form, while keeping it equal to its original value. To simplify 4/2, for example, you would change it to 2, because they are equal to one another. To simplify the equation $3n + 2n + 2 + 1$, you would combine the like terms and get $5n + 3$. Simplifying does not necessarily require a specific value to be obtained and should not be confused with solving.

slide: Moving an object a certain distance while maintaining the size and orientation (direction) of the object. This is also known as translation. Example: Scooting a book on a table. *See translation.*

slope: The amount of change in the *y*-coordinate with respect to the amount of change in the *x*-coordinate in a straight line on a graph. It is represented by the constant m in the equation $y = mx + b$. Slope can be found by taking any two points on a line, finding the difference in the *y* values, and then dividing that difference by the difference in the corresponding *x* values. The rise over the run.

solve: To find the solution to an equation or problem; finding the values of unknown variables that will make a true mathematical statement.

sphere: A closed surface consisting of all points in space that are the same distance from a given point (the center). Example: A basketball.

square: A rectangle with congruent sides. *See rectangle.*

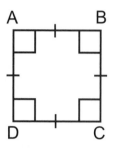

square root: The square root of a number A is the number which when multiplied by itself equals A. Example: 7 and -7 are square roots of 49 because 7 x 7 = 49 and (-7) x (-7) = 49. Every positive number has two square roots, one that is positive and one that is negative. The principal square root of a number (denoted $\sqrt{x}$) is its positive square root. Note the difference in the answers to these questions.

　　1. What is the square root of 81?
　　　　Answer: 9 and -9.
　　2. What is $\sqrt{81}$?
　　　　Answer: 9 only, not -9.

GLOSSARY

standard units of measure: Units of measure commonly used, generally classified in the U.S. as the customary system or the metric system

---

Customary System:
  Length
  1 foot (ft) = 12 inches (in)
  1 yard (yd) = 3 feet, or 36 inches
  1 mile (mi) = 1,760 yards, or 5,280 feet

  Weight
  16 ounces (oz) = 1 pound (lb)
  2,000 pounds = 1 ton (t)

  Capacity
  1 pint (pt) = 2 cups (c)
  1 quart (qt) = 2 pints
  1 gallon (gal) = 4 quarts

Metric System:
  Length
  1 centimeter (cm) = 10 millimeters (mm)
  1 decimeter (dm) = 10 centimeters
  1 meter (m) = 100 centimeters
  1 kilometer (km) = 1,000 meters

  Weight
  1,000 milligrams (mg) = 1 gram (g)
  1,000 grams (g) = 1 kilogram (kg)

  Capacity
  1 liter (L) = 1,000 milliliters (mL)

---

strategy: A plan used in problem solving, such as looking for a pattern, drawing a diagram, working backward, etc.

stem-and-leaf plot: A type of graph that depicts data by occurrence, using commonalities in place value. The digit in the tens place is used as the stem. The digit in the ones place is used as the leaf. Data is arranged like the example below.

Example: Ages of Adults in the Park

Data set				Stem	Leaves			
23	25	29	29	2	3	5	9	9
36	38	39	39	3	6	8	9	9
52	54	55	55	5	2	4	5	5

straight angle: An angle with a measure of 180°; this is also a straight line.

subtraction: An operation that removes sets from an initial set, or finds the difference between two amounts when comparing two quantities.

successive events: Events that follow one another in a compound probability setting.

sum: The result of addition. addend + addend = sum.

summary: A series of statements containing evidence, facts, and/or procedures that support a result.

supplementary angle: Two angles with a sum of 180°.

surface area: The sum of the areas of all of the faces (or surfaces) of a 3-D object. Also the area of a net of a 3-D object. Calculations of surface area are in square units (in^2, m^2, or cm^2).

GLOSSARY

survey: To get an overview by gathering data.

symbol: A letter or sign used to represent a number, function, variable, operation, quantity, or relationship. Examples: a, =, +, …

symmetrical: Having a line, plane, or point of symmetry such that for each point on the figure, there is a corresponding point that is the reflection of that point. *See line of symmetry.*

table: A method of displaying data in rows and columns.

tessellation: A pattern formed by placing congruent figures together with no empty space or overlapping areas. An example of a tessellation is a checkerboard.

three-dimensional figure: A shape (geometric figure) having length, width, and height.

translation: A transformation of a figure by sliding without turning or flipping in any direction. *See slide.*

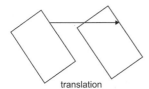

translation

trapezoid: Usually a trapezoid is defined as a quadrilateral that has exactly two parallel sides. Less often, it is defined as a quadrilateral with at least two parallel sides. (There is not complete agreement on the definition of a trapezoid.)

tree diagram: A visual diagram of all the possible outcomes for a certain event. A tree diagram is used to show the probability of a certain event happening.

trend: The general direction or tendency of a set of data.

triangle: The figure formed by joining three non-collinear points with straight segments. Some special types of triangles include equilateral, isosceles, and right triangles.The sum of the angles of a triangle is always equal to 180º.

turn: To move a point or figure in a circular path around a center point. Motion may be either clockwise or counterclockwise. Example: The hands of a clock turn around the center of the clock in a clockwise direction. *See rotation.*

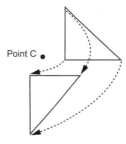

Point C

two-dimensional figure: A shape (geometric figure) having length and width. (A flat figure)

U.S. system of measurement: This is the system of measurement that most people in the United States use; the rest of the world, for the most part, uses the metric system. The following is a list of common U.S. units of measurement along with their abbreviations: length: inches, feet, yards, miles (in, ft, yd, mi); volume: fluid ounces, teaspoons, tablespoons, cups, pints, quarts, gallons (fl oz, tsp., tbsp.,c., pt., qt., gal.); weight: ounces, pounds, tons (oz, lb, t.); temperature: degrees Fahrenheit (ºF). Also called the customary system of measurement.

GLOSSARY

undefined terms: A term whose meaning is not defined in terms of other mathematical words, but instead is accepted with an intuitive understanding of what the term represents. The words "point," "line," and "plane" are undefined terms from geometry.

unknown: In algebra, the quantity represented by a variable.

variable: A symbol used to represent a quantity that changes or can have different values. Example: In 5*n*, the n is a variable.

verify: To establish as true by presentation of evidence.

vertex: In a two-dimensional object, any point where two segments join to form an angle. In a three-dimensional object, any point where three or more segments join to form a corner of the object. In a cube, for example, there are 8 vertices.

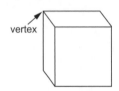

vertex

vertical angles: The pair of angles opposite to one another at the point where two lines, segments, or rays intersect. Vertical angles are always congruent to one another.

vertices: Plural of vertex.

volume: The amount of area taken up by a three-dimensional object is known as its volume. The units of measurement used to express volume can be cubic units, such as cubic feet or cubic centimeters, or, when measuring fluids, units such as gallons or liters. Volume is usually abbreviated as V and is also called capacity.

whole number: An integer in the set {0, 1, 2, 3 . . .}. In other words, a whole number is any number used when counting, in addition to zero.

word forms: The expression of numbers and/or symbols in words. Examples: 546 is "five hundred forty-six." The "<" symbol means "is less than." The ">" symbol means "is greater than." The "=" symbol means "equals" or "is equal to."

***x*-axis:** One of two intersecting straight (number) lines that determine a coordinate system in a plane; typically the horizontal axis.

***y*-axis:** One of two intersecting straight (number) lines that determine a coordinate system in a plane; typically the vertical axis.

zero property: This property states that in addition the sum of a given number and zero is equal to the given number. In multiplication, the product of zero and any number is zero. For example, any number *a*, in addition: $a + 0 = a$; in multiplication: $a \times 0 = 0$.

Mathematics Practice Tutorial

Directions for the Mathematics Practice Tutorial

The Mathematics Practice Tutorial contains 30 practice questions, a Mathematics Reference Sheet (from pages 33–34), and an Answer Sheet. You are permitted use of a calculator, and you will mark your answers on the Answer Sheet on pages 123–125 of this workbook. If you don't understand a question, just ask your teacher to explain it to you.

This section will review the Strands, Standards, and Benchmarks used to assess student achievement in the state of Florida. Following the description of each Benchmark, a student strategy to help you answer the question and a sample mathematics practice item is given. Each item gives you an idea of how the Benchmark may be assessed. Review these items to increase your familiarity with FCAT-style multiple-choice and gridded-response items. Once you have read through this tutorial section, you will be ready to complete the Mathematics Assessments.

Tips for Using a Calculator on the FCAT Mathematics Test

Here are some hints to help you show what you know when you take the Mathematics Practice Tutorial and the Mathematics Assessments:

- Read the problem carefully. Then decide whether or not you need the calculator to help you solve the problem.

- When starting a new problem, always clear your calculator by pressing the clear key.

- If you see an **E** in the display, clear the error before you begin.

- If you see an **M** in the display, clear the memory and the calculator before you begin.

- If the number in the display is not one of the answer choices, check your work. Remember that when computing with certain types of fractions, you may have to round the number in the display.

- Remember, your calculator will NOT automatically perform the algebraic order of operations.

- Calculators might display an incorrect answer if you press the keys too quickly. When working with calculators, use careful and deliberate keystrokes, and always remember to check your answer to make sure that it is reasonable.

- Always check your answer to make sure that you have completed all of the necessary steps.

Don't Forget...

Mathematics test questions with this symbol ✏️ require that you fill in a grid on your answer sheet. There may be more than one correct way to fill in a response grid. The gridded-response section on page 61 will show you the different ways the response grid may be completed.

Sample Multiple-Choice Item

To help you understand how to answer the test questions, look at the sample test question and Answer Sheet below. It is included to show you what a multiple-choice question in the test is like and how to mark your answer on your Answer Sheet.

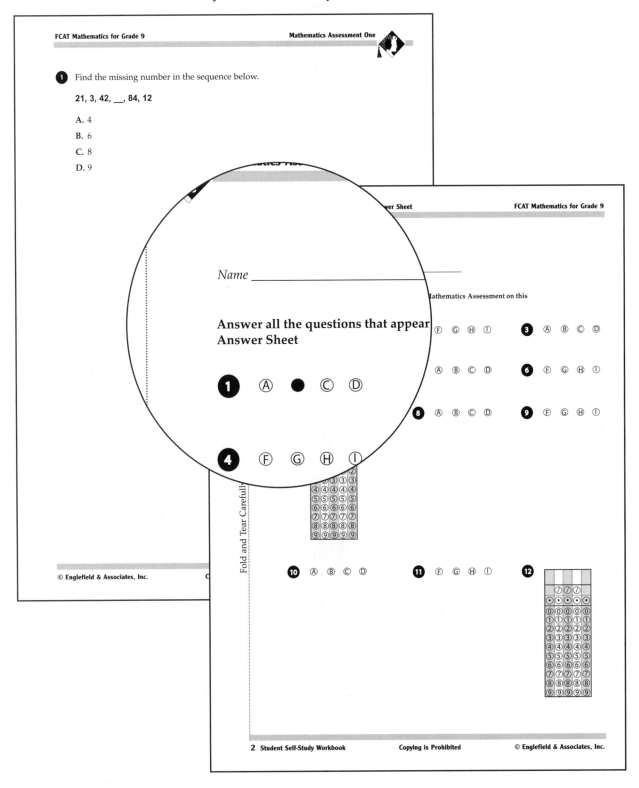

FCAT Mathematics for Grade 9   **Mathematics Assessment One**

1   Find the missing number in the sequence below.

21, 3, 42, __, 84, 12

A. 4
B. 6
C. 8
D. 9

© Englefield & Associates, Inc.

Name _____

Answer all the questions that appear
Answer Sheet

Mathematics Assessment on this

1   (A)   ●   (C)   (D)

(F)   (G)   (H)   (I)   **3**   (A)   (B)   (C)   (D)

(A)   (B)   (C)   (D)   **6**   (F)   (G)   (H)   (I)

8   (A)   (B)   (C)   (D)   **9**   (F)   (G)   (H)   (I)

4   (F)   (G)   (H)   (I)

Fold and Tear Carefully

10   (A)   (B)   (C)   (D)   **11**   (F)   (G)   (H)   (I)   **12**

2 **Student Self-Study Workbook**   Copying is Prohibited   © Englefield & Associates, Inc.

Sample Gridded-Response Item

To help you understand how to answer the test questions, look at the sample test question and Answer Sheet below. It is included to show you what a gridded-response item in the test is like and how to mark your answer on your Answer Sheet.

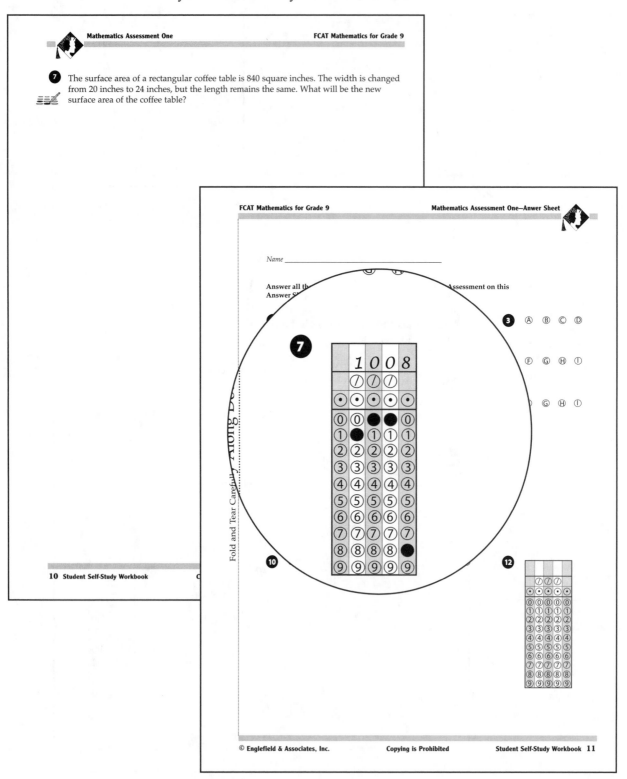

Additional Gridded-Response Item Information

Each gridded-response question requires a numerical answer which should be filled into a bubble grid. The bubble grid consists of 5 columns. Each column contains numbers 0–9 and a decimal point; the middle three columns contain a fraction bar as well. You do not need to include any commas for numbers greater than 999. When filling in your answer, only fill in one bubble per column. All gridded-response questions are constructed so the answer will fit into the grid. You can print your answer with the first digit in the left answer box, or with the last digit in the right answer box. Print only one digit or symbol in each answer box. Do not leave a blank box in the middle of an answer. Make sure you fill in a bubble under each box in which you wrote an answer and be sure to write your answer in the grid above the bubbles as well, in case clarification is needed. Answers can be given in whole number, fraction, or decimal form. For questions involving measurements, the unit of measure required for the answer will be provided for you. When a percent is required to answer a question, do NOT convert the percent to its decimal or fractional equivalent. Grid in the percent value without the % symbol. You may NOT write a mixed number such as $13\frac{1}{4}$ in the answer grid. If your answer is a mixed number, you must convert the answer to an improper fraction, such as $\frac{53}{4}$, or to a decimal number, such as 13.25. If you try to fill in $13\frac{1}{4}$, it will be read as $\frac{131}{4}$ and be counted as wrong. You will also be instructed when to round your answer in a particular way. Some example responses are given below.

Answer: 23,901   Answer: 26.5   Answer: 0.071   Answer: $\frac{3}{8}$

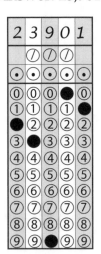

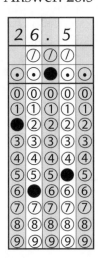

Question **1** *assesses*:

Strand A: Number Sense, Concepts, and Operations

Standard 1: The student understands the different ways numbers are represented and used in the real world.

MA.A.1.4.1 Associates verbal names, written word names, and standard numerals with integers, rational numbers, irrational numbers, real numbers, and complex numbers.

Student Strategies:

On the FCAT, you will be expected to explain to which of the following groups a given number belongs.

- **Counting Numbers or Natural Numbers:** 1, 2, 3, 4, 5, 6 and so on. The set of all positive integers.

- **Whole Numbers:** All of the Natural or Counting Numbers plus zero; 0, 1, 2, 3, 4, 5, 6 and so on.

- **Integers:** The set of all signed whole numbers; ..., -4, -3, -2, -1, 0, 1, 2, 3, 4, ...

- **Rational Numbers:** The set of all numbers that can be written as $\frac{a}{b}$, where both a and b are both integers. When these numbers are written as decimals, they either terminate (stop producing digits of their own accord) or repeat in some kind of pattern. These are examples of Rational Numbers: $\frac{3}{4}$, $\frac{13}{3}$, 5.8, 2.1717..., 5 (or any other whole number) because they can always be written over 1, e.g., $\frac{5}{1}$. The set of rational numbers includes all integers, whole numbers and natural numbers.

- **Irrational Numbers:** The set of all real numbers that cannot be written as $\frac{a}{b}$, where both a and b are integers. When these numbers are written as decimals, they neither terminate,(stop producing digits of their own accord) nor repeat in some kind of pattern. These are examples of irrational numbers: π = 3.14159265358979323..., e = 2.718281828459045..., $\sqrt{2}$, $\sqrt{3}$, or the square root of any number that's not a perfect square.

- **Real Numbers:** Includes all irrational numbers, rational numbers, integers, whole numbers and natural numbers.

- **Complex Numbers:** Have a real part and an imaginary part. Any number in the form $a + ib$, where a and b are both integers and i is the square root of -1.

1 Which of the following describes the number **7,641.211**?

A. seven million six hundred forty-one thousand two hundred eleven
B. seven thousand six hundred forty-one and two hundred elevenths
C. seven thousand six hundred forty-one and two hundred eleven thousandths
D. seven thousand six hundred forty-one and two hundred eleven

Go On ▶

Analysis: *Choice C is correct. Understanding the place value of numbers and how they are related through words is an important skill to master. The charts below diagram the place values of each digit in the given numbers. The fourth place left of the decimal is the thousands place, not the millions place, so Choice A is incorrect. This is no such thing as an elevenths place so Choice B is incorrect. There is a decimal point between the first 1 and the 2, so this part of the number must be less than one, not greater. Choice D is incorrect.*

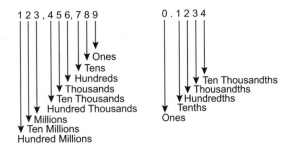

Question **2** *assesses*:

Strand A: Number Sense, Concepts, and Operations

Standard 1: The student understands the different ways numbers are represented and used in the real world.

MA.A.1.4.2 Understands the relative size of integers, rational numbers, irrational numbers, and real numbers.

Student Strategies:

It is not only important to be able to name and classify numbers, but also to be able to compare their relative sizes. In some instances, you may be asked to compare decimals to fractions, fractions to percents, or to compare other combinations of numbers in different formats. In these cases, you will want to convert the numbers into the same format.

Many students find the decimal format easiest. To compare decimals, it is necessary to consider the place values of each number, specifically, the "most significant digit." The most significant digit is the furthest left, non-zero digit in the number. It has the greatest value and is the most important determiner of size. Only "leading zeros," that is zeros in the furthest position *left of the decimal*, cannot be the most significant digit. For example, even though the number 0.1241 may seem greater than 0.2 based on appearance, it is not. The first digit in both of these numbers is not the most significant digit because it is a zero in both cases. The number in the tenths place takes this role. Since the number 0.2 has a larger significant digit in the tenths place than 0.1241 has, 0.2 is the greater of the two.

2 The career batting averages for five baseball players are given below. Which answer choice lists the players in order from **highest** batting average to **lowest** batting average?

Name	Average
Roberto Clemente	0.317
Jackie Robinson	0.311
Wade Boggs	0.328
Joe DiMaggio	0.325
George Brett	0.305

F. George Brett, Wade Boggs, Joe DiMaggio, Roberto Clemente, Jackie Robinson

G. George Brett, Jackie Robinson, Roberto Clemente, Joe DiMaggio, Wade Boggs

H. Wade Boggs, Roberto Clemente, Jackie Robinson, Joe DiMaggio, George Brett

I. Wade Boggs, Joe DiMaggio, Roberto Clemente, Jackie Robinson, George Brett

Go On ▶

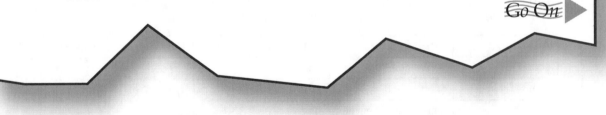

Analysis: Choice I is correct. In the question above, the numbers being compared are already in the same format; they are all decimals. Also, these numbers all have the same digit in the tenths place. This being the case, move to the next place, the hundredths place, and compare the numbers. If some of the numbers are still the same, move to the next place. Using this method, after comparing the hundredths place, you can see that George Brett has the lowest career batting average (0.305). Also, while Wade Boggs and Joe DiMaggio both have higher career batting averages than Roberto Clemente and Jackie Robinson (because of a higher number in the hundredths place), you must examine the numbers in the thousandths place to find the final order of career batting averages from greatest to least: 0.328, 0.325, 0.317, 0.311, 0.305, or Wade Boggs, Joe DiMaggio, Roberto Clemente, Jackie Robinson, George Brett. Choices F, G, and H are incorrect because they do not list the names in the correct order.

Question ❸ *assesses*:

Strand A: Number Sense, Concepts, and Operations

Standard 1: The student understands the different ways numbers are represented and used in the real world.

MA.A.1.4.3 Understands concrete and symbolic representations of real and complex numbers in real-world situations.

Student Strategies:

Learn to recognize key words found in word problems that indicate operations or symbols. The chart below summarizes the most common key words.

Addition	Subtraction	Multiplication	Division	Equals
added to	subtracted from	multiplied by	divided by	is
sum	minus	product of	quotient of	are
total of	difference of/between	times	per	was
more than	fewer than	increased or decreased by a factor of	ratio of	were
increased by	decreased by		out of	will be
combined together	less or less than	of	percent	gives
		double/multiply by 2	a (sometimes)	yields
		triple/multiply by 3		sold for

3 Kristie bought a new house. The yard had a few trees in it, but Kristie wanted to triple the number of trees in the yard. If *t* represents the number of trees in the yard before Kristie planted more trees, which of the following expressions represents the number of trees in the yard after she planted the new trees?

A. $t \times 3$
B. $t + 3$
C. $3(t + 1)$
D. $t \times 4 - 1$

Go On ▶

Analysis: Choice is A is correct. Pay attention to the details given to you through numbers and words. The question tells you that Kristie wanted to "triple" the number of trees in the yard. When you triple something, you multiply it by 3. In this situation, the thing being tripled is the number of trees, which is represented by the variable t. Since the number of trees is unknown, using a variable to represent it makes the expression work for any actual number of trees. The word triple implies multiplication, not addition, therefore, Choice B is incorrect. The only thing being tripled is the present number of trees, t. There is no reason to add anything to t before multiplying it by 3, therefore, Choice C is incorrect. Multiplication should be done before subtractions so the amount in Choice D is one less than four times t, not three times t. Choice D is incorrect.

Question **4** *assesses*:

Strand A: Number Sense, Concepts, and Operations

Standard 1: The student understands the different ways numbers are represented and used in the real world.

MA.A.1.4.4 Understands that numbers can be represented in a variety of equivalent forms, including integers, fractions, decimals, percents, scientific notation, exponents, radicals, absolute value, and logarithms.

Student Strategies:

When you are presented with a large variety of number formats, it is best to convert the numbers to decimals.

Convert a fraction to a decimal by dividing the numerator by the denominator.

A number written in scientific notation has two parts: a decimal fraction between 1 and 10 and some power of 10. Convert a number in scientific notation to standard notation (a standard decimal number) by first noticing the power of the ten. The power's sign, whether positive or negative, tells which way to move the decimal and the power's absolute value tells how many places to move it. If the power is positive you will move the decimal to the right. If the power is negative you will move the decimal to the left. For example, in 3.451×10^{-7}, notice that the power of the 10 is negative and that its absolute value is 7. This means that the decimal point has to move 7 places to the left. Since there is currently only one digit left of the decimal point, six zeros have to be added in front of the 3. Thus, $3.451 \times 10^{-7} = .0000003451$. If the 10 had had a positive power, such as 3.451×10^{7}, the decimal point would have to be moved 7 places to the right. So, $3.451 \times 10^{7} = 34,510,000$. Notice that since there are only three digits right of the decimal, four zeros have to be added after the 1.

Convert percents to decimals by dividing by 100. This moves the decimal two places to the left. If you get confused changing decimals to percents or percents to decimals, it is useful to use this little device. Write down a D for decimal and P for percent. Always write the letter D first since D comes before P in the alphabet. If you want to convert decimals to percents, draw an arrow from the D to the P as so: D→P. If you want to convert percents to decimals, draw an arrow from the P to the D as so: D◄—P. When converting from decimals to percents or from percents to decimals you **always** move the decimal point two places. The two diagrams above tell you which way to move it.

4 Which of the following numbers is NOT equivalent to the others?

F. 0.125

G. $\frac{19}{152}$

H. 1.25 x 10^1

I. 12.5%

Go On ▶

Analysis: *Choice H is correct. Change all other number formats to decimals. Since Choice F is already a decimal and since it is the same as Choices G and I, Choice F is incorrect. In Choice G, 19 ÷ 152 = 0.125. Choice G is incorrect. In Choice H, 1.25 x 10^1 = 1.25 x 10 = 12.5. Since this is not the same as 0.125, Choice H is correct. In Choice I, 12.5% ÷ 100 = 0.125. Choice I is incorrect.*

Question **5** *assesses*:

Strand A: Number Sense, Concepts, and Operations

Standard 2: The student understands number systems.

MA.A.2.4.2 Understands and uses the real number system.

Student Strategies:

Some problems use two or more different units to measure the same type of quantity. You will usually have to convert these to like units. For example, if speed is listed as miles per hour in one instance and feet per second in another, one of these units will have to be converted into the other.

5 On average, a beaver cuts down 200 trees per year. How many trees per month does a beaver cut down on average?

　　A. 20 trees per month
　　B. 16.7 trees per month
　　C. 13.3 trees per month
　　D. 10 trees per month

Go On ▶

Analysis: *Choice B is correct. To find the number of trees cut down by a beaver in a month, take the total number of trees cut down by a beaver in a year and divide it by 12, the number of months in a year: 200 trees ÷ 12 months ≈ 16.7 trees per month. Choice A is incorrect because 20 trees per month times 12 months equals 240 trees per year, not 200. Choice C is incorrect because 13.3 trees per month times 12 months equals about 160 trees per year, not 200. Choice D is incorrect because 10 trees per month times 12 months equals 120 trees per year, not 200.*

Question **6** *assesses:*

Strand A: Number Sense, Concepts, and Operations

Standard 3: The student understands the effects of operations on numbers and the relationships among these operations, selects appropriate operations, and computes for problem solving.

MA.A.3.4.1 Understands and explains the effects of addition, subtraction, multiplication, and division on real numbers, including square roots, exponents, and appropriate inverse relationships.

Student Strategies:

Squaring a number means multiplying that number by itself. Another way to think of squaring a number is raising the number to the second power. Specific examples include:

1. Squaring any number, whether positive or negative, always results in a positive number. For example, $7^2 = 49$ and $(-7)^2 = 49$.

2. Squaring any fraction between 0 and 1 always results in a smaller number.

   First square the numerator, then square the denominator. For example,

$$(\tfrac{1}{4})^2 = \tfrac{1}{4} \times \tfrac{1}{4} = \tfrac{1}{16} \text{ or } (\tfrac{1}{4})^2 = \tfrac{1^2}{4^2} = \tfrac{1}{16}$$

3. Squaring the square root of any number results in the number itself. For example, $(\sqrt{11})^2 = 11$

4. Squaring two numbers and then multiplying them gives the same answer as multiplying them first and then squaring the result. For example, $(3 \times 6)^2 = 18^2 = 324$ and $(3 \times 6)^2 = 3^2 \times 6^2 = 9 \times 36 = 324$.

5. A negative exponent does not mean a negative number, it means a reciprocal. So, a number raised to the -2 power means the reciprocal of the number squared. For example, $5^{-2} = \tfrac{1}{5^2} = \tfrac{1}{25}$

6. When squaring a square, multiply powers to get a base to the fourth power.

   For example, $(5^2)^2 = 5^4 = 625$ or $(5^2)^2 = 25^2 = 625$

6 Which of the following numbers results in the **smallest** value when squared?

F. 2^{-2}

G. $3\sqrt{2}$

H. $\frac{1}{3}$

I. -3

Go On ▶

Analysis: Choice F is correct. To find the correct answer, square all the choices and then compare their

sizes. In Choice F, $(2^{-2})^2 = 2^{-4} = \frac{1}{2^4} = \frac{1}{16}$. In Choice G, $(3\sqrt{2})^2 = 3^2(\sqrt{2})^2 = 9 \times 2 = 18$. This is larger

than $\frac{1}{16}$, so Choice G is incorrect. In Choice H, $(\frac{1}{3})^2 = \frac{1}{3} \times \frac{1}{3} = \frac{1}{9}$. This is larger than $\frac{1}{16}$, so Choice H

is incorrect. In Choice I, $(-3)^2 = 9$. This is larger than $\frac{1}{16}$, so Choice I is incorrect.

Question **7** *assesses*:

Strand A: Number Sense, Concepts, and Operations

Standard 3: The student understands the effects of operations on numbers and the relationships among these operations, selects appropriate operations, and computes for problem solving.

MA.A.3.4.2 Selects and justifies alternative strategies, such as using properties of numbers, including inverse, identity, distributive, associative, and transitive, that allow operational shortcuts for computational procedures in real-world or mathematical problems.

Student Strategies:

Be careful with variables in formulas. Variables may be single letters, subscripted letters, or even whole words. The same letters with two different subscripts are really two completely different variables. For example, x_1 and x_2 are just as much different variables as a and b. Subscripted variables are often used when two values are different, but somehow related. An upper-case letter and a lower-case letter of the same letter are also two different variables. For example, A and a are different variables which may or may not be related. Make sure you substitute the correct value into each variable.

7 A group of scientists was studying avalanches. They came up with a formula that could be used to determine the volume of snow in cubic feet per second, (V_t) moving in an avalanche after the avalanche had been moving for a certain time period (t_t) measured in seconds. Their equation, where V_i represents the initial volume of snow moving in cubic feet and t represents the total duration of the avalanche in seconds, is shown below. They witnessed an avalanche that began with 15 cubic feet of snow and lasted 27 seconds. How much snow was moving in the avalanche after 20 seconds? Round your answer to the nearest tenth. Your answer will be in **cubic feet** per second.

Avalanche Equation: $V_t = \dfrac{V_i \times (t_t)^2}{t}$

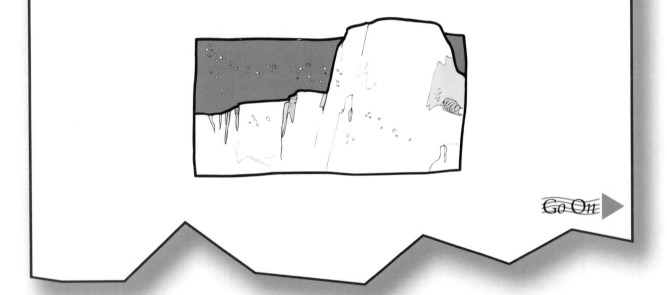

Go On ▶

Analysis: The correct response is 222.2 cubic feet per second. Substitute the proper values into the equation and solve:

$$V_t = \frac{(15)(20^2)}{27} = \frac{(15)(400)}{27} = \frac{6000}{27} = 222.2 \text{ cubic feet per second}$$

Question **8** *assesses*:

Strand A: Number Sense, Concepts, and Operations

Standard 3: The student understands the effects of operations on numbers and the relationships among these operations, selects appropriate operations, and computes for problem solving.

> **MA.A.3.4.3** Adds, subtracts, multiplies, and divides real numbers, including square roots and exponents, using appropriate methods of computing, such as mental mathematics, paper and pencil, and calculator.

Student Strategies:

Not all information provided in a problem is equally useful. Read the problem carefully to distinguish between vital information and extraneous or irrelevant information. Occasionally you may be given a problem that is unsolvable as stated because insufficient information is provided.

8 As a fund-raising project, the 12 members of the ski club agree to clean the football stadium after each home football game for a fee of $200.00. The football stadium contains a total of 744 benches. Approximately how much money does the ski club earn per bench cleaned?

A. $0.02
B. $0.17
C. $0.27
D. $0.62

Go On ▶

Analysis: Choice C is correct. To find the correct answer, be sure you understand what the question is asking you. The question is worded "earn per bench." The word "per" gives you a clue that something must be divided by something else. The word "earn" has to do with money, in this case, the total amount of money earned by the ski club members. This total will be divided by the total number of benches, which satisfies the "bench" aspect of the question. Therefore, the ski club earns $200.00 ÷ 744 benches ≈ $0.27 per bench cleaned. The number of ski club members is not important in this question. Choices A, B, and D are incorrect because they are either guesses or result from an incorrect division.

Question **9** *assesses*:

Strand A: Number Sense, Concepts, and Operations

Standard 4: The student uses estimation in problem solving and computation.

MA.A.4.4.1 Uses estimation strategies in complex situations to predict results and to check the reasonableness of results.

Student Strategies:

There are many reasons to estimate instead of calculating an exact answer. Among these are:

1. You may not need an exact answer. You may only need to determine a value in a rough or general way.

2. You may not want to invest the time, effort, or expense necessary to calculate an exact answer.

3. It may be impossible to calculate an exact answer because the data you have is incomplete or imperfect.

4. You may need to use some sort of sampling or polling technique because the data you are trying to measure keeps changing. In the time it would take to get an exact answer, the data changes so much that conclusions drawn from it are useless.

Be aware that if a question on the FCAT calls for an estimate and you provide an exact answer, you may lose points.

9 Dana and Lorraine are taking a trip to Chelan, Washington. The trip to Chelan is 2,289.9 miles from their starting point. If their car can go 27 miles per gallon of gas, ESTIMATE the number of gallons of gas they will need to get to Chelan.

F. 38 gallons
G. 54 gallons
H. 77 gallons
I. 154 gallons

Go On ▶

Analysis: *Choice H is correct. The method used to solve this problem is rounding. Begin by rounding the number of miles: 2,289.9 miles is rounded to 2,300 miles. Next, round the number of miles the car can travel using one gallon of gas: 27 miles per gallon rounds to 30 miles per gallon. Now, divide the rounded number of miles by the rounded miles per gallon to find the estimated number of gallons of gas they will need to get to Chelan: 2,300 miles ÷ 30 miles per gallon = 76.6 gallons or 77 gallons. Choices F, G, and I are incorrect because they are not close estimates. Note that this estimated answer varies somewhat from the value you get if you divide 2289.9 by 27: 2289.9 ÷27 ≈ 85 gallons. This answer is about 10% larger than our first estimate, however Choice H is still the closest answer choice. Different estimates will produce different answers. The user of the estimate needs to determine how precise the estimate should be. Also, remember, the 27 miles per gallon is also an estimate.*

Question **10** *assesses:*

Strand B: Measurement

Standard 1: The student measures quantities in the real world and uses the measures to solve problems.

MA.B.1.4.1 Uses concrete and graphic models to derive formulas for finding perimeter, area, surface area, circumference, and volume of two- and three-dimensional shapes, including rectangular solids, cylinders, cones, and pyramids.

Student Strategies:

Be very careful what you take for granted from a diagram or sketch accompanying a problem. There are very few things you can assume from the way a sketch looks. Making unwarranted assumptions can cause you to get incorrect answers.

What you can assume:

- Lines that appear straight are straight.
- Lines that appear to intersect, do intersect.
- Points that appear to be on a line or curve are on the line or curve.
- Anything stated as a given, or that can be logically deduced from a given using sound logic or known theorems.
- Polygons that appear closed are closed.

What you cannot assume:

Anything about the relative size of anything measurable. Examples of mistaken assumptions in geometric drawings.

- The "length" of a rectangle is not required to be longer than its "width."
- One angle may appear larger than another when it isn't.
- One line segment may appear longer than another when it isn't.
- Angles may appear acute, obtuse, right, or congruent when they are not.
- Lines may appear parallel or perpendicular when they are not.
- Triangles may appear acute, obtuse, right, equilateral, isosceles, or scalene when they are not.
- Polygons may appear regular, irregular, similar, or congruent when they are not.
- Points not on a line may appear to be collinear when they are not.
- A point appears to be the midpoint of a line segment when it is not.

10 Mr. Duncan has a stone wall made of rectangular stones around his garden. Each spring, he replaces the stones that have fallen out with new stones. What is the perimeter of the stone that will fit in the hole in the stone wall shown below?

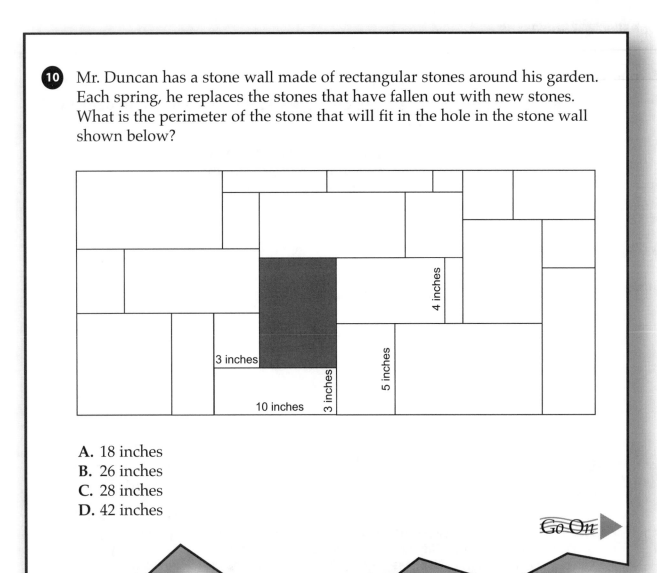

A. 18 inches
B. 26 inches
C. 28 inches
D. 42 inches

Go On ▶

Analysis: Choice B is correct. Because there are no direct measurements given for the dimensions of the hole, you must use the information given about other stones to determine the height and the width of the hole. The width of the hole can be found using the dimensions of Stone D (see diagram below). Stone D has dimensions of 10 inches by 3 inches. You can also see that part of the top width of Stone D is 3 inches because the width of Stone A makes up part of Stone D's width. This means the remaining length of the top of Stone D is 7 inches. This distance is the same as the width of the hole. The height can be found in a similar fashion. Part of the height of the hole is 4 inches because part of the height of the hole is composed of the height of Stone B, which is given as 4 inches. The rest of the height can be found using the height of Stone D and the height of Stone C. The height of Stone C is 5 inches. The length of the height on the left side of Stone C is split into two parts. One part is the same as the height of Stone D, which is 3 inches. This means the remaining section of the height of Stone C is 2 inches. Therefore, the total height of the hole is 6 inches. Now, to find the perimeter, add together the length of each side of the hole, keeping in mind to use the height and width twice each: 7 inches + 7 inches + 6 inches + 6 inches = 26 inches. Note that the drawing is not to scale and the 7-inch "width" of the hole appears shorter than the 6-inch "length."

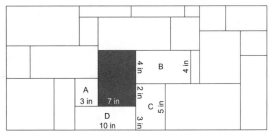

Question **11** *assesses:*

Strand B: Measurement

Standard 1: The student measures quantities in the real world and uses the measures to solve problems.

MA.B.1.4.2 Uses concrete and graphic models to derive formulas for finding rate, distance, time, angle measures, and arc lengths.

Student Strategies:

Never introduce more variables than you need into a solution. Always try to look for relationships between the different variables in your formula and try to identify a "key" value or parameter. A "key" value is used by the author to define other values. If possible, express other values in terms of this "key," rather than introducing new variables.

Example: A strip of paper 33 inches long is cut into three pieces. The longest piece is 8 inches longer than the middle-sized piece, and the shortest piece is 5 inches shorter than the middle-sized piece. How long are the three pieces? Solution: Since the question asks the length of the three pieces, some students may start out with three variables, s, m, and l, for short, middle-sized, and long. If this approach is taken, it is impossible to solve the problem except by trial and error. A better strategy is to concentrate on the key unknown, that is, the length that defines the other lengths. In this example, it is the middle-sized piece. If it is expressed as m, then the longest piece can be expressed as $m + 8$ and the shortest piece can be expressed as $m - 5$. Since together they equal 33 inches: $(m - 5) + m + (m + 8) = 33$; $3m + 3 = 33$; $3m = 30$; $m = 10$. So, $m - 5 = 5$ and $m + 8 = 18$. Thus, the three lengths are 5 in, 10 in, and 18 in.

11 Mary and Bruce are running along the same straight path toward each other. Mary is running about 15 miles per hour, and Bruce is running at 18 miles per hour. If they are 5 miles apart on the path, approximately how long will it be before they are at the same point on the path?

F. 15 minutes
G. 9 minutes
H. 7.5 minutes
I. 2.27 minutes

Go On ▶

Analysis: Choice G is correct. Use the distance formula, d = rt, or distance = rate x time. Since Bruce is running faster than Mary, he will cover a greater distance in the same amount of time. However, both people will run the same amount of time before they meet. Since we now know that the time, t, is the same for both runners, divide both sides of the distance formula by t to get: $\frac{d}{r}$ = t. If we take the distance Mary runs and divide it by her rate, we will get a fraction that we can set equal to a similar fraction about Bruce, the distance he runs divided by his rate. We can set them equal because we know that the time, t, is the same for both runners. Before we set these two fractions equal, however, we need to think about the distance, d, for both runners. It's clear that d can't be the same for both of them. Bruce's distance has to be larger than Mary's because he is running faster. We don't know what either d is, but we do know that it is somewhere between 0 and 5 since the runners are 5 miles apart and they meet somewhere in between. If we leave Bruce's d as d, we can replace Mary's with 5 – d because we know that Mary's distance plus Bruce's distance must add up to 5, the distance they are apart. We are given how fast each runner is moving, so substitute values for the variables to get each runner's fraction, their $\frac{d}{r}$. Bruce's fraction is $\frac{d}{18}$ while Mary's fraction is $\frac{5-d}{15}$. Setting these equal to each other yields: $\frac{d}{18} = \frac{5-d}{15}$. Cross-multiplying we get, 15d = 18(5 – d); 15d = 90 – 18d; 33d = 90; d ≈ 2.73 miles. Putting this distance into Bruce's equation: t = $\frac{d}{r}$; t ≈ $\frac{2.73\ miles}{18\ miles\ per\ hour}$ ≈ 0.15 hours. We can convert this fraction of an hour to minutes by multiplying it by 60 since there are 60 minutes per hour: 0.15 x 60 ≈ 9 minutes.

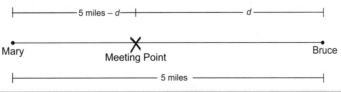

Question **12** *assesses:*

Strand B: Measurement

Standard 1: The student measures quantities in the real world and uses the measures to solve problems.

MA.B.1.4.3 Relates the concepts of measurement to similarity and proportionality in real-world situations.

Student Strategies:

Proportions and similarity are often used to measure things indirectly. This means that an item's properties are not measured directly with a ruler, scale, or measuring cup, but by comparison to the known dimensions of another related item. A proportion is simply an equation with two equivalent ratios. Think of it as setting two equivalent fractions equal to each other. For example, you know that 50/100 is the same as 1/2, so the proportion would be 1/2 = 50/100. Since each fraction has a numerator and a denominator, there are four possible pieces of information. You must know at least three of them. The fourth is the unknown and is represented as a variable. There are several ways a proportion can be set up correctly and also several ways to set it up incorrectly. The important thing to remember is to keep the information in the two fractions in the same order. Suppose you have a problem involving a tree's height, a man's height, and the length of both of their shadows. If one fraction is: $\frac{\text{height of tree}}{\text{length of tree's shadow}}$, the other fraction has to be: $\frac{\text{height of man}}{\text{length of man's shadow}}$. The second fraction cannot be flipped upside down. There are several other ways to correctly represent this proportion. Two more are listed below:

$$\frac{\text{length of tree's shadow}}{\text{height of tree}} = \frac{\text{length of man's shadow}}{\text{height of man}}$$

$$\frac{\text{length of tree's shadow}}{\text{length of man's shadow}} = \frac{\text{height of tree}}{\text{height of man}}$$

12 For a history project, Kristin is constructing a scale model of the Leaning Tower of Pisa. The actual height of the tower is 55.86 m and the actual diameter of the base is 4.93 m. If the base of Kristin's model has a diameter of 8 cm, approximately how tall should her model be? Your answer should be in **centimeters** rounded to the nearest hundredth.

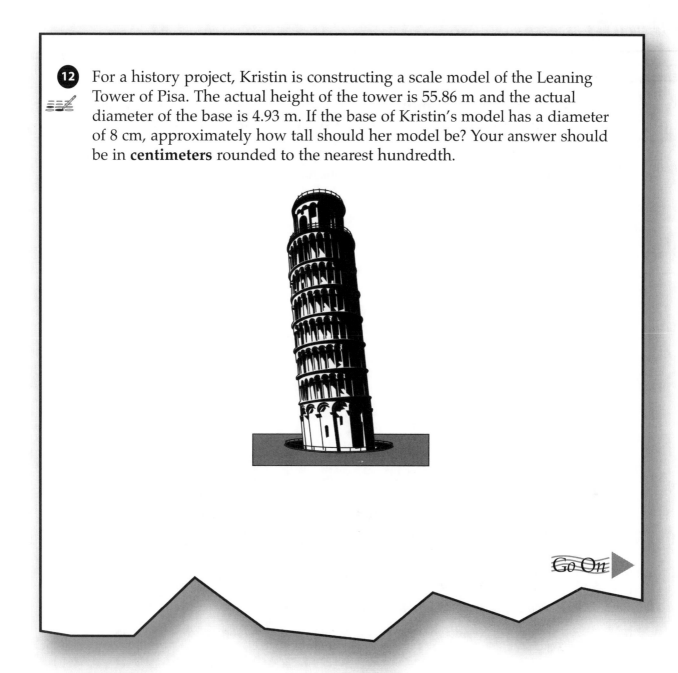

Go On ▶

Analysis: *The correct response is approximately 90.65 cm. To solve this question, you must set up a proportion. In this question, the ratios are the height of the tower to the diameter of its base is equal to the height of the model to the diameter of its base. Using given data, this can be written as follows:*

$$\frac{55.86 \text{ m}}{4.93 \text{ m}} = \frac{n \text{ (unknown height of model)}}{8 \text{ cm}}$$

To solve the proportion, cross-multiply: (55.86)(8) = (n)(4.93). This gives you 446.88 = 4.93n. Divide both sides by 4.93 to find n: n = 90.65 cm.

Question **13** *assesses:*

Strand B: Measurement

Standard 2: The student compares, contrasts, and converts within systems of measurement (both standard/nonstandard and metric/customary).

MA.B.2.4.1 Selects and uses direct (measured) or indirect (not measured) methods of measurement as appropriate.

Student Strategies:

Many real-world problems ask you to calculate how many of a certain item you need to perform a given task. Examples include:

How many gallons of paint do you need to paint the house?

How many dozen eggs do you need to make cookies for 200 people?

How many boxes of tile do you need to cover the floor?

How many buses do you need to transport the entire freshman class to the art gallery?

Many times the number of units you need to complete your task may not be a whole number, so you will have to buy a bit more than you need if the items required are sold in certain fixed quantities. For example, if you have to transport 113 people on buses that accommodate 45 passengers each, you will have to order three buses, not two and a half. It's necessary to order more than you need since you can't complete the task on less. Some problems on the FCAT may contain closely related answers to test whether you understand this principle.

Copying is Prohibited

© Englefield & Associates, Inc.

13 The 407 students in the ninth-grade class at Coughlan High School are having a class picnic. Debby is in charge of ordering pizzas for the picnic. She is told to get enough pizza so each student can have 3 pieces of pizza. If each pizza contains 8 pieces of pizza, how many pizzas does Debby need to order?

A. 51 pizzas
B. 136 pizzas
C. 152 pizzas
D. 153 pizzas

Go On ▶

Analysis: Choice D is correct. If each student can have 3 pieces of pizza, then multiplying the total number of students by 3 will give the total number of pieces of pizza needed: 407 students x 3 pieces per student = 1,221 pieces. Now, to find the number of pizzas, divide the total number of pieces by the number of pieces per pizza: 1,221 pieces ÷ 8 pieces per pizza = 152.625 pizzas. You must round this up because a whole pizza must be ordered; 0.625 of a pizza cannot be ordered. Therefore, the number of pizzas Debby must order is 153. Choice A is incorrect because it represents the number of pizzas necessary to give each student only one piece (407 ÷ 8 ≈ 51). Choice B is incorrect because this answer is obtained by dividing 407 by 3 instead of multiplying 407 by 3 to find the total number of pieces needed. Choice C is incorrect because it would be 5 pieces short of the required amount.

Question **14** *assesses*:

Strand B: Measurement

Standard 2: The student compares, contrasts, and converts within systems of measurement (both standard/nonstandard and metric/customary).

MA.B.2.4.2 Solves real-world problems involving rated measures (miles per hour, feet per second).

Student Strategies:

It is often necessary to convert the data in a problem from one type of unit to another. Care must be taken to do this correctly. The most frequently used units of speed are: miles per hour; feet per second; kilometers per hour; centimeters per second; and revolutions per minute. Notice that all of these speed units have the word "per" in them to imply a division. Conversion from one of these units to another usually entails a division, a multiplication, a proportion, or a combination of these. Speed expressed as revolutions per minute (rpm) can be particularly tricky. It might seem that any point on a spinning CD is traveling at the same speed, but this is not so. It is true that all points are turning at the same number of revolutions per minute, but those points further away from the center are on the circumferences of increasingly larger circles. Since these points are covering a greater distance with each revolution, they are moving faster than points closer to the center. To find the speed of a particular point as a linear distance per unit of time, e.g., miles per hour, feet per second, or inches per minute, you must find the circumference of the point's circle and multiply that by the number of revolutions per minute (or other unit of time). For example, Earth has an equatorial circumference of 24,900 miles and it revolves on its axis once every 24 hours, so 24,900 miles ÷ 24 hours ≈ 1,038 miles per hour. This means that a person "standing still" on the equator is really moving around Earth's axis at over 1,000 miles per hour.

---

14 Before CDs, people played music on record players with grooved vinyl discs called records. A record is spinning on a record player at 45 rpm (revolutions per minute). Given the measurements in the diagram below, how much faster is Point A traveling than Point B?

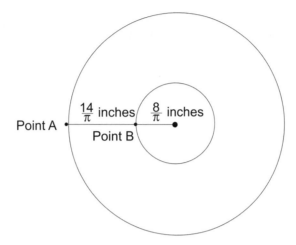

    F. 401 inches per minute
    G. 630 inches per minute
    H. 1,260 inches per minute
    I. 1,980 inches per minute

Go On ▶

Analysis: Choice H is correct. To find the difference in speed between Point A and Point B, you must first find the speed of Point A and Point B. Both points are turning at 45 rpm, but Point A is on the circumference of a bigger circle, so it's covering more distance with each revolution; it is moving faster than Point B. To find the distance that each point travels in one revolution, find the circumference of the circle it is on. From the drawing we can see that the distance from the center of the circle to Point B is $\frac{8}{\pi}$ inches, which is also Point B's radius. The distance from Point B to Point A is $\frac{14}{\pi}$ inches, so the radius of Point A's circle is $\frac{8}{\pi} + \frac{14}{\pi} = \frac{22}{\pi}$. Applying the formula for circumference, Point A travels $C = 2\pi r$; $C = 2\pi(\frac{22}{\pi})$; $C = \frac{2\pi}{1}(\frac{22}{\pi}) = 44$ inches. Point B travels $C = 2\pi r$; $C = 2\pi(\frac{8}{\pi})$; $C = \frac{2\pi}{1}(\frac{8}{\pi}) = 16$ inches. Since Point A travels 44 inches during each of its 45 revolutions per minute, it is traveling $44 \times 45 = 1,980$ inches per minute. Similarly, Point B travels 16 inches during each of its 45 revolutions per minute, or $16 \times 45 = 720$ inches per minute. The difference between these two speeds is: $1,980 - 720 = 1,260$ inches per minute.

Question **15** *assesses*:

Strand B: Measurement

Standard 3: The student estimates measurements in real-world problem situations.

MA.B.3.4.1 Solves real-world and mathematical problems involving estimates of measurements, including length, time, weight/mass, temperature, money, perimeter, area, and volume, and estimates the effects of measurement errors on calculations.

Student Strategies:

When you estimate, you usually need to round off one or more numbers. If all the numbers in your set are relatively close to each other in size, you usually round them to the same level of precision. Numbers are considered relatively close when they are of the same order of magnitude, which means that the largest number in the set is no more than 10 times larger than the smallest number. For example, if you have the numbers 6.29, 9.7, 1.9, 3.45, you might round all of them to the nearest whole number: 6, 10, 2, 3. In this case, if the digit in the tenths position is 5 or greater, round the whole number up. If the digit in the tenths position is 4 or less, round the whole number down. When rounding numbers of vastly unequal size, we often round them to the "most significant digit." The most significant digit is the left most, non-zero digit in the number. This digit has the greatest value in the number. As an example, compare the sun's radius to Earth's. Earth's radius is 3,959 miles. The most significant digit is the 3 in 3,959 and the number next to 3 is a 9, so we round our most significant digit up and call Earth's radius about 4,000 miles. The sun's radius is 432,500 miles. The 4 is the most significant digit and the digit next to it is 3, so we round down and call the sun's radius about 400,000 miles. Therefore, the sun's radius is 100 times greater than Earth's: 400,000 ÷ 4,000 = 100. On the FCAT, finding an exact answer to a problem when you are asked to estimate may lose you points.

15 Evelyn is filling the fish tank below with water. So far, she has filled half the tank with water. ESTIMATE the amount of water Evelyn will need to add to fill the tank to the line shown.

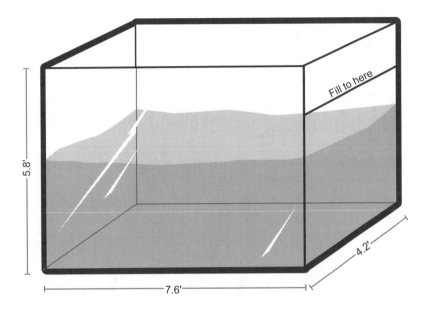

A. 35 cubic feet
B. 48 cubic feet
C. 60 cubic feet
D. 96 cubic feet

Go On ▶

Analysis: Choice B is correct. Begin by rounding the measurements to the nearest whole number to find the fish tank's estimated dimensions: 4 feet x 8 feet x 6 feet. So, the estimated volume of the entire tank is 4 feet x 8 feet x 6 feet = 192 cubic feet. The question tells you Evelyn has already filled half the tank with water, meaning half the tank is also empty. The line which indicates how high the water should be is in the middle of the space which has not yet been filled. This means that of the half which is not yet filled, only half of it needs to be filled. Half of a half is a fourth of the whole, so a fourth of the volume of the entire tank needs to be added: 192 cubic feet ÷ 4 = 48 cubic feet. Choice A is incorrect because it results from incorrectly rounding all three tank dimensions down before dividing by 4. Choice C is incorrect because it results from incorrectly rounding all 3 tank dimensions up before dividing by 4. Choice D is incorrect because it represents one-half the tanks volume, not one-fourth.

Question **16** *assesses*:

Strand C: Geometry and Spatial Sense

Standard 1: The student describes, draws, identifies, and analyzes two- and three-dimensional shapes.

MA.C.1.4.1 Uses properties and relationships of geometric shapes to construct formal and informal proofs.

Student Strategies:

Make sure you know the definitions of different quadrilaterals. The Venn diagram shows how the properties of each type relate to the others.

Definitions of Quadrilaterals

Quadrilateral: a closed four-sided polygon

Parallelogram: a quadrilateral with both pairs of opposite sides parallel

Rhombus: a quadrilateral with four congruent sides

Rectangle: a quadrilateral with four right angles

Square: a quadrilateral with four right angles and four congruent sides

Trapezoid: a quadrilateral with exactly one pair of parallel sides

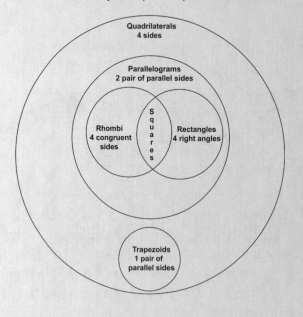

16 The quadrilateral below is a square. Which of the following is NOT true?

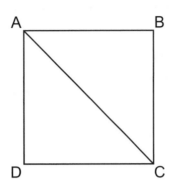

   F. ABCD is also a rhombus.
   G. The sum of the interior angles of ABCD is 360°.
   H. ABCD is also a trapezoid.
   I. Triangle ABC is an isosceles triangle.

Go On ▶

Analysis: Choice H is correct. By definition, a trapezoid has only one pair of opposite sides which are parallel to one another; a square has two pairs of parallel sides. Choice F is incorrect because every square is also a rhombus. Choice G is incorrect because all quadrilaterals including squares have a total interior angle sum of 360°. Choice I is incorrect because triangle ABC must be isosceles. Since all squares have four equal sides AB = BC, but these are also the two legs of the triangle.

Question **17** *assesses:*

Strand C: Geometry and Spatial Sense

Standard 2:The student visualizes and illustrates ways in which shapes can be combined, subdivided, and changed.

MA.C.2.4.1 Understands geometric concepts such as perpendicularity, parallelism, tangency, congruency, similarity, reflections, symmetry, and transformations including flips, slides, turns, enlargements, rotations, and fractals.

Student Strategies:

Any two polygons are similar if, and only if, their corresponding angles are congruent and their corresponding sides are proportional. This means that you must be very careful to match up corresponding angles and corresponding sides. Also, the scale factor which is the ratio of the lengths of two corresponding sides must be the same for all other pairs of corresponding sides. For example, consider the two similar quadrilaterals below. When you refer to them by their vertex letters, you must list corresponding points in the same order. If you refer to Figure ABCD, you must refer to the other quadrilateral as Figure EFGH, because A corresponds to E, B corresponds to F, and so on. The angles of these figures are shown to be congruent because angles A and E both have one arc, angles B and F both have two arcs, and so on. Congruent angles are indicated with the same number of arcs. Likewise, if any sides are equal, that is indicated with the same number of short strokes through corresponding line segments. Finally, the ratios of all the corresponding sides must be equal. For example:

$$\frac{AB}{EF} = \frac{BC}{FG} = \frac{CD}{GH} = \frac{AD}{EH}$$

This will also be the ratio of their perimeters, but not their areas.

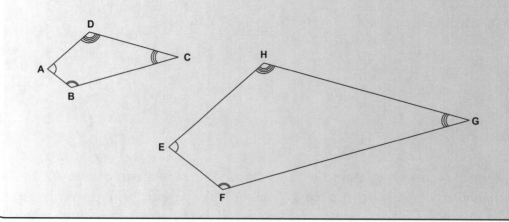

17 The triangles pictured below are similar. What is the perimeter of Triangle RST? Round your to the **hundredths** place.

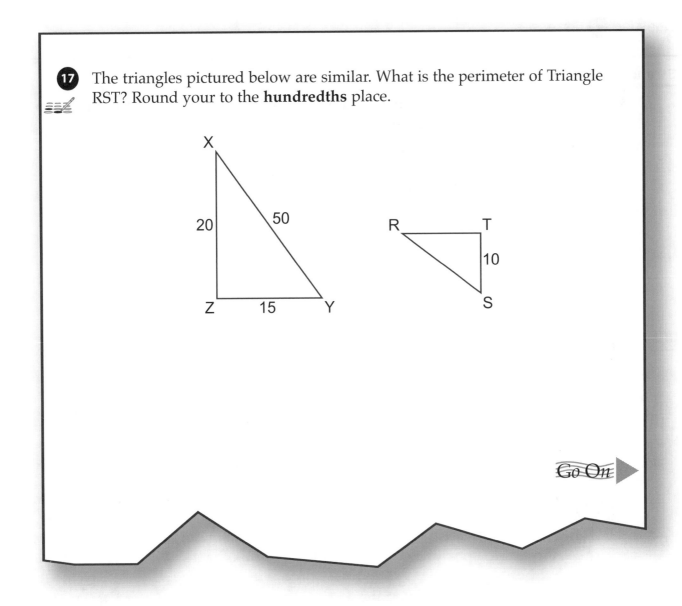

Go On ▶

Analysis: The correct response is 56.67. To find the perimeter, you must first find the correct relationship between the two triangles. Because the triangles are similar, they will have the same angle measurements and the corresponding sides will be related by a direct proportion. YZ corresponds to ST; the ratio of ST to YZ is 10 to 15 which reduces to 2 to 3; you can use this ratio to find the lengths of all the missing sides in Triangle RST and then add them together to find the perimeter of Triangle RST. Or, you could find the perimeter of Triangle XYZ and multiply it by the same ratio to find the perimeter of Triangle RST. The perimeter of Triangle XYZ is 85, so the perimeter of Triangle RST is 85 x 2/3 ≈ 56.67.

Question **18** *assesses*:

Strand C: Geometry and Spatial Sense

Standard 3: The student uses coordinate geometry to locate objects in both two and three dimensions and to describe objects algebraically.

MA.C.3.4.1 Represents and applies geometric properties and relationships to solve real-world and mathematical problems including ratio, proportion, and properties of right triangle trigonometry.

Student Strategies:

Probably the best way to learn about rotations (or any other transformation) is to overlay the graph of the object with a blank overhead projector transparency. To start, the graph of your shape should have three to five points with integer coordinates, named with letters of the alphabet. Trace the shape onto the plastic and stick a push-pin through the transparency and the graph at the point of rotation. Questions on the FCAT may ask you to rotate your object through multiples of 90°, so practice rotating the shape through 90°, 180°, and 270°. Be sure to keep your point of rotation fixed, but notice where all the other points end up after each rotation. Record the coordinates of these points both before and after the rotation and see if you can make a general statement about how they change. Try the experiment several times with different shapes.

18 The line in the graph below represents the equation $y = x + 3$. If the line were rotated 180° around Point A, what would be the new coordinates of Point B?

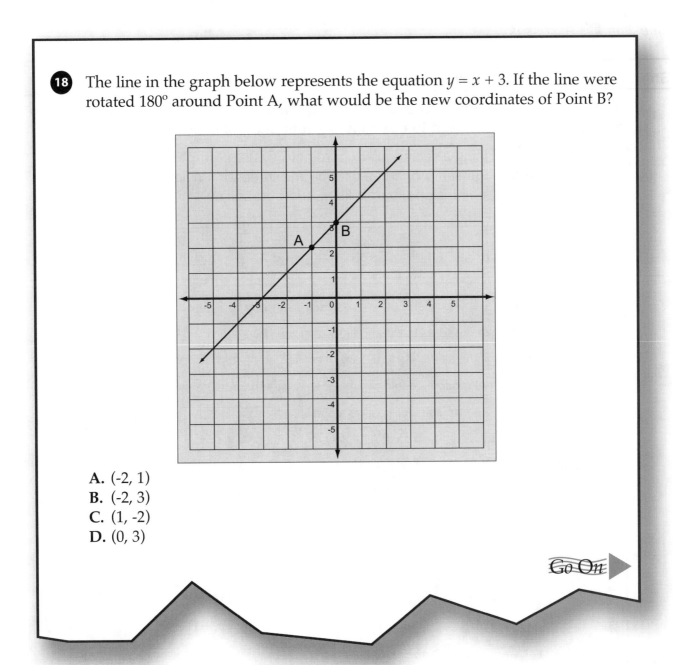

A. (-2, 1)
B. (-2, 3)
C. (1, -2)
D. (0, 3)

Go On ▶

Analysis: Choice A is correct. Rotating the line 180° around Point A does not change the line itself in any way; only the location of Point B on the line changes. In fact, the line's equation remains $y = x + 3$ after the rotation. Point B will remain the same distance away from Point A, but it will be located on the other side of Point A. Instead of being one coordinate up and one coordinate to the right of Point A, Point B will be one coordinate down and one coordinate to the left of Point A. The new coordinates of Point B are therefore (-2, 1). Choice B is incorrect because it represents Point B's location after a rotation of 270°, not 180°. Choice C is incorrect because it mixes up the correct values of -2 and 1 and puts them in the wrong order using x's value for y and y's value for x. Choice D is incorrect because it represents B's location after a 360° rotation or after no rotation at all.

Question **19** *assesses*:

Strand C: Geometry and Spatial Sense

Standard 3: The student uses coordinate geometry to locate objects in both two and three dimensions and to describe objects algebraically.

MA.C.3.4.2 Using a rectangular coordinate system (graph), applies and algebraically verifies properties of two- and three-dimensional figures, including distance, midpoint, slope, parallelism, and perpendicularity.

Student Strategies:

The slope of a line is its "slant." It is often expressed as the line's rise over its run: $\frac{\text{rise}}{\text{run}}$. Slope can also be expressed as the change in y over the change in x: $\frac{\Delta y}{\Delta x} = \frac{y_2 - y_1}{x_2 - x_1}$. A straight line maintains the same "slant" or slope between **any** two points on it. If you pick any point on the line at random, then pick any other point and calculate the slope, you will get the same value. An increasing or positive slope will show the line going up as you move from left to right along the x-axis. A decreasing or negative slope will show the line going down as you move from left to right along the x-axis. A horizontal line has a slope of zero. Slope is undefined in a vertical line, that is, a vertical line has no slope at all. The four possibilities for slope are illustrated below.

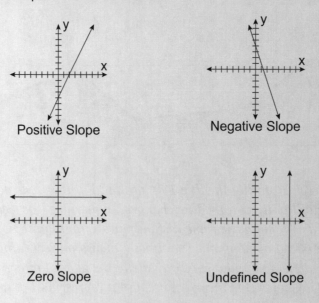

Positive Slope

Negative Slope

Zero Slope

Undefined Slope

19 Look at the graph below. Determine which of the following coordinates forms a line with a slope of 2 when connected to Point A.

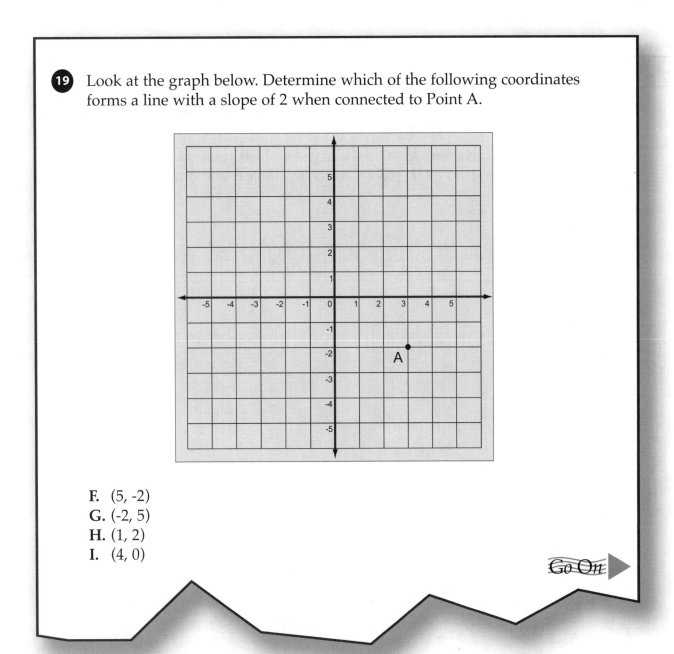

F. (5, -2)
G. (-2, 5)
H. (1, 2)
I. (4, 0)

Go On

Analysis: Choice I is correct. Slope is a line's "rise over run" or "the number of units moved in a vertical direction divided by the number of units moved in a horizontal direction." To find a line that has a slope of 2 and includes Point A, (3, -2), move up 2 units from Point A and then over to the right 1 unit (when dealing with slope, up and right are "positive moves," while down and left are "negative moves"). The "rise" is 2, and the "run" is 1. The coordinate up 2 units and right 1 unit from Point A is (4, 0). Had this not been an answer choice, you could go up another 2 units and right another 1 unit to find point (5, 2). Similarly, you could also work backward from Point A to find (2, -4). Choice F is incorrect because a line containing pints (3, -2) and (5,-2) has a slope of 0, not a slope of 2. Choice G is incorrect because a line containing points (3, -2) and (-2, 5) has a slope of -7/5, not -2. Choice H is incorrect because a line containing points (3, -2) and (1, 2) has a slope of -2, not 2.

Question **20** *assesses:*

Strand D: Algebraic Thinking

Standard 1: The student describes, analyzes, and generalizes a wide variety of patterns, relations, and functions.

MA.D.1.4.1 Describes, analyzes, and generalizes relationships, patterns, and functions, using words, symbols, variables, tables, and graphs.

Student Strategies:

Sometimes when you are asked to translate a written problem into a mathematical expression, it helps to make a table summarizing what you know and what you can figure out. For test item question 20 on the next page, the table might look like this:

Number of Drinks (*n*)	1	2	3	4	5
Total Cost (*c*)	$1.95	$2.70	$3.45	$4.20	$4.95

The correct equation will give each of these predicted *c*-values when the corresponding *n*-value is substituted into it.

20 A snack stand on the beach has the following special: if you buy a souvenir mug for $1.95, you can have it refilled for $0.75. If n equals the number of times the mug is filled and c equals the total cost, which of the following equations shows how to determine the cost for any number of times the mug is filled?

A. $(n - 1)\$1.95 + \$0.75n = c$
B. $\$1.95 + \$0.75n = c$
C. $\$1.95 + (n - 1)\$0.75 = c$
D. $(n - 1)(\$1.95 + \$0.75) = c$

Go On ▶

Analysis: Choice C is correct. First, you need to include a variable for cost in the equation because that is what the question asks you to find. This is the "$= c$" in each of the answer choices. Next, you need to determine what makes up the total cost. The original cost of the souvenir mug is the first thing to consider. This is shown in each answer choice by "$\$1.95$." The total cost of any number of refills is the second thing to consider. This is represented in the equation by "$(n - 1)\$0.75$." The expression $\$0.75n$ is not correct because with the first purchase, the customer is charged only for the cup, not the drink in it. Using $(n - 1)$ eliminates the "refill charge" on the first fill-up because when $n = 1$, the expression becomes 0, and if the refill charge is multiplied by 0, it will be 0. To complete the equation, put together the two parts of the total cost using the correct operation. In this case, the operation is addition: $\$1.95 + (n - 1)\$0.75 = c$. Choices A, B, and D are incorrect because none of them will yield the correct value of c when a given value of n is entered. In fact, none of them even give the correct value for the first drink.

Question **21** *assesses:*

Strand D: Algebraic Thinking

Standard 1: The student describes, analyzes, and generalizes a wide variety of patterns, relations, and functions.

MA.D.1.4.2 Determines the impact when changing parameters of given functions.

Student Strategies:

Sometimes it seems as if there are a million formulas for you to remember and apply, but if you try to understand organizing mathematical concepts, you can cut down on what you have to memorize and you will better understand problem solving. For example, the volume of any cylinder or any right prism (triangular, rectangular, hexagonal, etc), can be found with the formula: $V = Bh$ where B is the area of the base and h is the height of the object. For example, consider the triangular prism below:

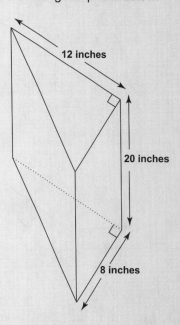

12 inches

20 inches

8 inches

To find the volume of this prism, all we have to do is find the area of one of its triangular bases and multiply that by its height. The area of this triangle is: $A = \frac{1}{2}bh = \frac{1}{2}(8)(12) = \frac{1}{2}(96) = 48$ in². Since the height of the prism is 20 inches, its volume is: $V = Bh = 48 \times 20 = 960$ in³. Make sure you notice that there are two heights in this problem. The 12 inches is the height of the triangular base while the 20 inches is the height of the prism.

21 At a beverage factory, a can of juice is completely filled at a constant rate in 4 seconds. The dimensions of the can of juice are shown in the diagram below. How much more juice is in the can after 3 seconds of filling than after 1 second of filling? Your answer will be in **cubic inches**. Round your answer to the nearest whole number.

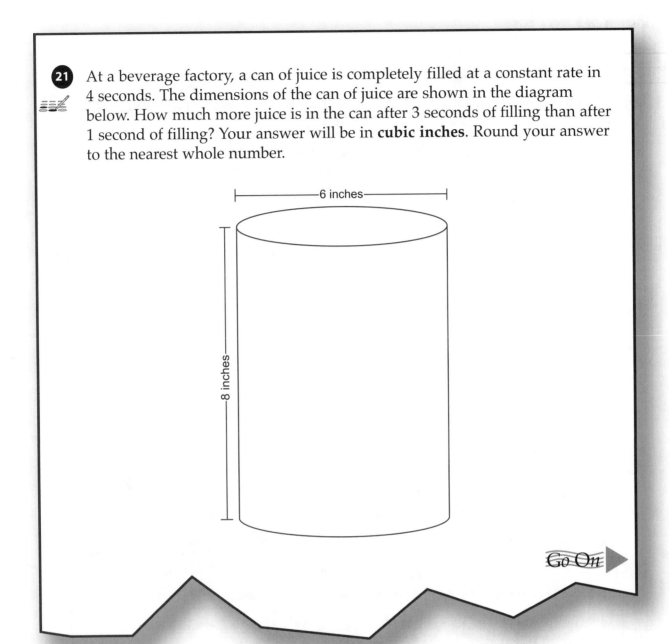

Analysis: *The correct response is about 113 cubic inches (answers may vary slightly depending on the value of π used). The formula for the volume of a cylinder is: V = πr²h. Notice that this formula fits the pattern of V = Bh, where B is the area of the base and h is the height of the cylinder. The base of a cylinder is a circle and the area of a circle is found with πr². To answer this question we'll need to find the volume of the can at two different heights, but the area of the base will remain the same in each case. A circle with a diameter of 6 inches has a radius of 3 inches, so the area of the base is: A = πr² = π × 3² = 9 π. Now we must determine the height of the juice in the can after 1 second and after 3 seconds of filling. Since we're told that the can is completely filled at a constant rate in 4 seconds, the can must be 1/4 full after 1 second, 1/2 full after 2 seconds, 3/4 full after 3 seconds, and completely full (4/4) after four seconds. The height of the can is 8 inches, so after 1 second it will be 1/4 full and juice will be 8 inches × 1/4 = 2 inches high in the can. After 3 seconds the can will be 3/4 full and juice will be 8 inches × 3/4 = 6 inches high in the can. The volume of juice after 1 second will be: V = Bh = 9π × 2 = 18π in³. The volume of juice after 3 seconds will be: V = Bh = 9π × 6 = 54π in³. This means the difference will be: 54π cubic inches − 18π cubic inches = 36π cubic inches ≈ 113 cubic inches.*

Question **22** *assesses*:

Strand D: Algebraic Thinking

Standard 2: The student uses expressions, equations, inequalities, graphs, and formulas to represent and interpret situations.

MA.D.2.4.1 Represents real-world problem situations using finite graphs, matrices, sequences, series, and recursive relations.

Student Strategies:

Whenever you're asked to graph a linear equation or match a linear equation with its graph, put the equation in $y = mx + b$ form, if possible. The b in this equation is the y-intercept of the line. This is the point at which the graph of the equation crosses the y-axis. Plot or look for this point first. Its coordinates will always be $(0, b)$. The m in this equation is the slope of the line. This tells how steeply the line rises or falls and is interpreted as the rise over the run, $\frac{rise}{run}$, or as the change in y over the change in x, $\frac{\Delta y}{\Delta x} = \frac{y_2 - y_1}{x_2 - x_1}$. Always think of this number m as a fraction. If it's a whole number, put it over 1 to make it a fraction. If the slope is negative, just put the negative sign in the numerator. Starting at point $(0, b)$, the y-intercept, go up or down the number of units in the rise then go right the number of units in the run and plot another point. Draw a line through these two points and use the slope again in the same way as before to plot a third point to check your work.

Copying is Prohibited
© **Englefield & Associates, Inc.**

22 Which of the following graphs represents the equation $y = \frac{2}{3}x - 4$?

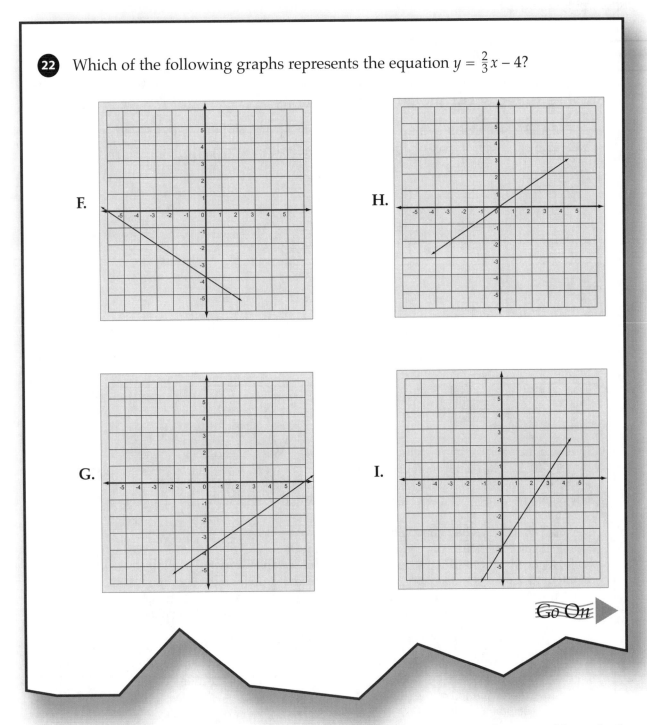

Go On ▶

Analysis: Choice G is correct. A good place to begin this problem is to see what y would equal when x is 0: $y = 2/3(0) - 4$ or $y = -4$. The correct graph will cross the y-axis at (0, -4). You can eliminate Choice H because the line does not contain a point at (0, -4). The slope of the equation is 2/3. By looking at the slope of the line, Choice F can also be eliminated because a line with a positive slope goes up from left to right; the line in Choice F does not. To determine which of the remaining choices is correct, try substituting another value for x into the equation. If x = 3, then y = -2. The line in Choice I does not contain this coordinate; Choice G does.

Question **23** *assesses:*

Strand D: Algebraic Thinking

Standard 2: The student uses expressions, equations, inequalities, graphs, and formulas to represent and interpret situations.

MA.D.2.4.2 Uses systems of equations and inequalities to solve real-world problems graphically, algebraically, and with matrices.

Student Strategies:

Formulas are valuable in mathematics and other disciplines because they generalize a relationship that exists between parameters, forces, and concepts. Formulas allow us to predict what will happen under certain circumstances. Any given formula or expression may contain a variety of letters called variables. Variables hold the value of a given parameter, standing in for this force or concept, until a specific number is substituted into it. Almost all formulas have one variable alone on one side of the equation while everything else is on the opposite side. This lone variable is the dependent variable. Its value *depends* on what is substituted into the other variables, the independent variables, on the other side of the equation. You will often be called upon to change formulas around. That is, you will be asked to make the dependent variable independent and one of the independent variables dependent. This is done by using three main concepts.

1. Isolate the one variable that you are asked to solve for.

2. Use "opposite operations" to undo your expression one piece at a time and help you isolate the variable. For example, if the formula uses addition, use subtraction to undo it. If the formula uses multiplication, use division or a reciprocal to undo it.

3. Whatever you do to one side of the equation, you must do to the other. Often, you will have to use two or more "opposite operations."

23 The equation $(F - 32) \times \frac{5}{9} = C$ is used to convert temperatures from degrees Fahrenheit to degrees Celsius temperatures (F = degrees Fahrenheit, C = degrees Celsius). Which of the following is an equivalent equation that solves for **degrees Fahrenheit**?

A. $\frac{9}{5} \times C + 32 = F$

B. $(C + 32) \times \frac{5}{9} = F$

C. $\frac{9}{5}(C + 32) = F$

D. $(C - 32) \div \frac{5}{9} = F$

Go On ▶

Analysis: Choice A is correct. To solve this problem, you must isolate F on one side of the equation. Begin by multiplying each side of the equation by 9. This gives you: $(F - 32) \times 5 = C \times 9$. Next, divide each side of the equation by 5. This leads to: $F - 32 = (C \times 9) \div 5$. Finally, add 32 to each side: $F = ((C \times 9) \div 5) + 32$. The "$\times 9 \div 5$" can also be written as "$\times 9/5$." If you are comfortable with fractions, a shorter solution method is to multiply both sides of the equation by the reciprocal of the fraction 5/9 and then add 32 to both sides:

$$(F - 32) \times \tfrac{5}{9} = C; \ (F - 32) \times \tfrac{5}{9} \times \tfrac{9}{5} = C \times \tfrac{9}{5}; \ F - 32 = \tfrac{9}{5}C; \ F - 32 + 32 = \tfrac{9}{5}C + 32; \ F = \tfrac{9}{5}C + 32$$

Choice B is incorrect because the variable C should be multiplied by the reciprocal of 5/9, not 5/9, then added to 32 instead of doing the addition first. Choice C is also incorrect because the addition is done before the multiplication. Choice D is incorrect because it subtracts 32 from the variable C instead of adding it. The division by 5/9 is the same as multiplying by 9/5, however.

Question **24** *assesses*:

Strand E: Data Analysis and Probability

Standard 1: The student understands and uses the tools of data analysis for managing information.

MA.E.1.4.1 Interprets data that have been collected, organized, and displayed in charts, tables, and plots.

Student Strategies:

Problems with graphs will often ask you to interpret data or combine it in new ways. You may also be asked to look for patterns and relationships, draw or justify conclusions, answer related questions, or compare characteristics of different graphs and data sets. To do this successfully, pay particular attention to the labels on graphs. Make sure you understand what is being measured and what units are being used. Remember that the person who created the graph wants to communicate some idea with it. Try to judge if the data is being represented accurately and fairly. For instance, ask yourself, "Does the scale start at an appropriate number? Are the measurement units appropriately sized to display important data without distorting it?" Often, when the graph represents very large numbers, to improve readability, the scale along the side shows rather small integers like 10, 20, 30, etc. with a parenthetical note near them like (in millions). These numbers are meant to represent 10 million, 20 million, 30 million, and so on.

24 According to the bar graph below, what number of glockenspiels did Chimea export from 1999–2001?

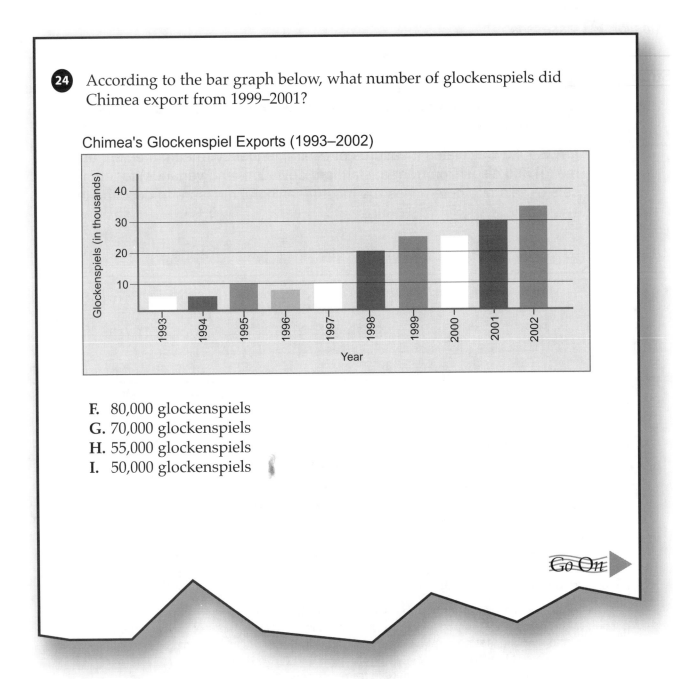

Chimea's Glockenspiel Exports (1993–2002)

F. 80,000 glockenspiels
G. 70,000 glockenspiels
H. 55,000 glockenspiels
I. 50,000 glockenspiels

Go On ▶

Analysis: Choice F is correct. Use the data in the graph to determine the correct answer. The bar for the number of glockenspiels exported in 1999 represents 25,000 glockenspiels. The bar for 2000 also represents 25,000 glockenspiels. The bar for 2001 represents 30,000 glockenspiels. Add these numbers together to find the total number of glockenspiel exports from 1999 to 2001: 25,000 glockenspiels + 25,000 glockenspiels + 30,000 glockenspiels = 80,000 glockenspiels. Choices G, H, and I are incorrect because of an incorrect reading of the chart units or an incorrect choice of export years.

Question **25** *assesses:*

Strand E: Data Analysis and Probability

Standard 1: The student understands and uses the tools of data analysis for managing information.

MA.E.1.4.2 Calculates measures of central tendency (mean, median, and mode) and dispersion (range, standard deviation, and variance) for complex sets of data and determines the most meaningful measure to describe the data.

Student Strategies:

Make sure you understand the differences between the three measures of central tendency, how to calculate each, and certain important characteristics of each.

The **mean** is the arithmetic average. To find the mean, add up all of the numbers in your data set and divide the total by the number of terms in the set. The mean is sensitive to extremes. These values, also called outliers, are either very much higher or very much lower than the rest of the members of the data set. Outliers can skew the mean, that is, make it look larger or smaller than the majority of scores would indicate. This is especially true if a few outliers exist at one end of the data set and are unbalanced by similar outliers on the other end of the data set.

The **median** is the middle term of the data set if there is an odd number of terms in the set, or the average of the two middle terms if there is an even number of terms in the set. To find the median, first arrange all the members of the data set in ascending numerical order, then count forward and backward from each end of the data set "one for you, one for me" fashion until you reach the middle term or terms. The median is not affected by outliers or extreme values in the data. It is possible for changes in individual values to change the mean of the set without affecting the median. The median is affected only by the relative order of the terms.

The **mode** of a data set is its most frequent value. It is unaffected either by outliers or the relative order of the terms. It is only affected by how many times each term occurs. It is possible to have more than one mode or no modes at all in a data set.

25 The table below shows the top 10 career goal scorers for the National Hockey League. What is the **mean** number of goals scored by these players?

Name	Goals
Wayne Gretzky	894
Gordie Howe	801
Marcel Dionne	731
Phil Esposito	717
Mike Gartner	708
Mark Messier	651
Brett Hull	649
Mario Lemieux	648
Steve Yzerman	645
Bobby Hull	610

Go On ▶

Analysis: The correct response is 705.4 goals. The mean of a set of data is the average of the values in the set. You can calculate mean by adding the values in the set together and then dividing by the total number of items in the set. For example, in this set, the number of goals is added together to find these individuals have scored a total of 7,054 goals in their careers. In the set, there are 10 items (items in this example can be thought of as the goal totals of each person). So, to find the mean, divide 7,054 goals by 10: 7,054 goals ÷ 10 people = 705.4 goals.

Question **26** *assesses*:

Strand E: Data Analysis and Probability

Standard 1: The student understands and uses the tools of data analysis for managing information.

MA.E.1.4.3 Analyzes real-world data and makes predictions of larger populations by applying formulas to calculate measures of central tendency and dispersion using the sample population data, and using appropriate technology, including calculators and computers.

Student Strategies:

One important application of statistics involves using random sampling to make an indirect measurement of something that is difficult or impossible to measure directly. This method, which is often used in a wildlife census such as in question 26, is really an estimate based on probability concepts, but if proper care is taken, can be fairly accurate. The idea is to obtain two sample populations. The first sample, which should be large enough to adequately represent the population, is obtained randomly. All members of the entire population should have an equal chance of being selected. Each individual in this sample is marked in some obvious way and released back into the general environment. These individuals must be allowed to mix freely with the general unmarked population because this technique depends upon dispersion or the completely random mixing of marked and unmarked individuals. The marked individuals are used as a sampling device to set up a proportion that will allow an estimate of the total number of individuals in the population. Next, a second sample should be chosen randomly. This sample should, by probability, include marked individuals in the same proportion as those in the entire population. In other words, this sample should be a scale model of the entire population. A properly set-up proportion should now yield an accurate estimate of the total population.

$$\frac{\text{number of marked individuals in second sample}}{\text{total number in second sample}} = \frac{\text{number of individuals marked in first sample}}{\text{total population size}}$$

---

26 A wildlife expert is trying to determine the number of squirrels that inhabit a forest. He catches 50 squirrels, tags them with markers he can identify, and then releases them. Two days later, he returns to the forest and catches 50 more squirrels. Of those squirrels, 12 of them have tags on them. Using this information, which of the following is a good prediction of the total number of squirrels in the forest?

A. 62 squirrels
B. 208 squirrels
C. 380 squirrels
D. 600 squirrels

Go On ▶

Analysis: *Choice B is correct. To make an accurate estimate, you must use both sample populations. The second time the wildlife expert caught 50 squirrels, 12 of them were tagged. If all squirrels have an equal chance of being sampled, the proportion of tagged squirrels in this sample should be equal to the proportion of tagged squirrels in the forest. Since we know there are 50 tagged squirrels in the forest (from the first 50 squirrels caught and released by the wildlife expert), we can use this proportion:*

$$\frac{50 \ (number \ of \ tagged \ squirrels \ )}{n \ (squirrels \ in \ the \ total \ population)} = \frac{12 \ (number \ of \ tagged \ squirrels \ in \ second \ sample)}{50 \ (total \ squirrels \ in \ second \ sample \ )}$$

Use cross-multiplication to solve for n: $12n = 2500$; $n \approx 208$. So there are about 208 squirrels in the forest. This is only an estimate, though. To find the actual number, you would have to count each individual squirrel. To obtain a more refined estimate, this process should be repeated several times, and the results should be compared and averaged. For this situation, though, 208 squirrels seems like a good estimate based on the given information.

Question **27** *assesses:*

Strand E: Data Analysis and Probability

Standard 2: The student identifies patterns and makes predictions from an orderly display of data using concepts of probability and statistics.

MA.E.2.4.1 Determines probabilities using counting procedures, table, tree diagrams, and formulas for permutations and combinations.

Student Strategies:

When most people think of combinations, they are really thinking of two distinct mathematical concepts—combinations and permutations. A **combination** is a selection of items where order does not matter. A **permutation** is a selection of items where order is vitally important. To contrast these concepts, imagine three people (Joe, Jane, and John) are selected to be members of a committee. It doesn't matter whether Joe or one of the other people is selected first. It only matters that each was selected to be a committee member. This is an example of a combination. Now suppose that Joe, Jane, and John are running for class office in an election where the highest vote-getter is elected president, the second highest is elected vice-president, and the third-highest is elected secretary. Here the order of the selection is vitally important. This is an example of a permutation.

27 At an ice cream parlor, ice cream cones are ordered by flavor, cone, and topping. The choices for flavors are chocolate, vanilla, and strawberry. The choices for cones are regular, sugar, and waffle. The choices for toppings are plain, sprinkles, or nuts. If one of each category must be used for each ice cream cone, how many different combinations of ice cream cones can be ordered?

F. 9 combinations
G. 18 combinations
H. 27 combinations
I. 54 combinations

Analysis: Choice H is correct. One way to determine the total number of combinations is to take the number of possible choices for each category and multiply them together: 3 flavors x 3 cones x 3 toppings = 27 combinations. You might also want to make a table of combinations to verify your results. Either way, the total number of possible combinations is 27.

chocolate, regular, plain	vanilla, regular, plain	strawberry, regular, plain
chocolate, regular, sprinkles	vanilla, regular, sprinkles	strawberry, regular, sprinkles
chocolate, regular, nuts	vanilla, regular, nuts	strawberry, regular, nuts
chocolate, sugar, plain	vanilla, sugar, plain	strawberry, sugar, plain
chocolate, sugar, sprinkles	vanilla, sugar, sprinkles	strawberry, sugar, sprinkles
chocolate, sugar, nuts	vanilla, sugar, nuts	strawberry, sugar, nuts
chocolate, waffle, plain	vanilla, waffle, plain	strawberry, waffle, plain
chocolate, waffle, sprinkles	vanilla, waffle, sprinkles	strawberry, waffle, sprinkles
chocolate, waffle, nuts	vanilla, waffle, nuts	strawberry, waffle, nuts

Question **28** *assesses*:

Strand E: Data Analysis and Probability

Standard 2: The student identifies patterns and makes predictions from an orderly display of data using concepts of probability and statistics.

MA.E.2.4.2 Determines the probability for simple and compound events as well as independent and dependent events.

Student Strategies:

Sample space is a very important concept in probability calculations and experiments. The sample space is simply the set of all possible outcomes of an experiment or event. A list or table of the complete sample space can be a very handy tool in complicated probability situations. Consider the sample space of rolling two standard dice. The table below is a complete list of all possible outcomes. It is the sample space for this situation and can be used to help solve question 28 on the next page.

		Second Die					
		1	2	3	4	5	6
First Die	1	1,1	1,2	1,3	1,4	1,5	1,6
	2	2,1	2,2	2,3	2,4	2,5	2,6
	3	3,1	3,2	3,3	3,4	3,5	3,6
	4	4,1	4,2	4,3	4,4	4,5	4,6
	5	5,1	5,2	5,3	5,4	5,5	5,6
	6	6,1	6,2	6,3	6,4	6,5	6,6

Go On

---

28 If two standard 6-sided dice are rolled, one after another, what is the probability that the second number rolled will be greater than the first number rolled?

A. $\frac{1}{6}$

B. $\frac{5}{12}$

C. $\frac{1}{2}$

D. $\frac{7}{12}$

Go On

Analysis: *Choice B is correct. First construct a sample space. There are 6 possible numbers on the first die and 6 possible numbers on the second die. This means that the total number of possible outcomes is 36 (6 x 6 = 36). Now determine how many of those combinations satisfy the question's requirements. They are outlined on the sample space below. For example, if a "1" is rolled first, then there are 5 possible numbers that could be rolled on the second die that are greater than 1: 2, 3, 4, 5, and 6. If a "2" is rolled, then there are 4 possible numbers that could be rolled on the second die that are greater than 2: 3, 4, 5, and 6. Continuing this type of counting shows that there are 15 combinations satisfying the question. To find the probability, divide the number of desired outcomes by the total number of possible outcomes: 15 desired outcomes/36 possible outcomes which reduces to 5/12.*

		Second Die					
		1	2	3	4	5	6
First Die	1	1, 1	1, 2	1, 3	1, 4	1, 5	1, 6
	2	2, 1	2, 2	2, 3	2, 4	2, 5	2, 6
	3	3, 1	3, 2	3, 3	3, 4	3, 5	3, 6
	4	4, 1	4, 2	4, 3	4, 4	4, 5	4, 6
	5	5, 1	5, 2	5, 3	5, 4	5, 5	5, 6
	6	6, 1	6, 2	6, 3	6, 4	6, 5	6, 6

15 combinations where the second number is larger than the first.

Question **29** *assesses:*

Strand E: Data Analysis and Probability

Standard 3: The student uses statistical methods to make inferences and valid arguments about real-world situations.

MA.E.3.4.1 Designs and performs real-world statistical experiments that involve more than one variable, then analyzes results and reports findings.

Student Strategies:

There are many kinds of graphs and charts and each type has its own strengths, weaknesses, and uses. A data set may be represented by a number of different types of graphs depending on what aspect of the data is being emphasized. It's useful to know some of the characteristics and uses of the most common types of graphs.

Characteristics and Uses of Common Types of Graphs	
Pictograph	
Remarks	**Uses**
Uses cartoons or simple drawings for quantities. Requires a legend to show the quantity represented by each unit.	Makes abstract comparisons more concrete. Good for young or unsophisticated audiences. Best used when there are large differences in data.
Bar Graph	
Remarks	**Uses**
Quantities shown as bars of varying length. Data sets must be relatively small.	Shows comparisons of same class over time. Comparisions of two or more classes possible. Time shown in discrete units (non-continuous). Useful to show ranked data.
Line Graph	
Remarks	**Uses**
An increasing/decreasing line shows changes over time. Shows only one comparison per line. Several lines may appear on the same axes.	Shows comparisons over time. Time shown as continuous. Used to find and compare trends. Can interpolate between data points. Can extrapolate beyond know values (forecast).
Pie Chart	
Remarks	**Uses**
Easy to understand. Not useful to show changes over time. Not useful to show exact comparisons of values, since estimating angles is difficult for many people. Not useful to show rank data.	Shows proportional relationships at one point in time. Shows a division of the whole or a part of the whole.

29 Use the graphs below to determine how many women in Townsville think Thanksgiving is the best holiday.

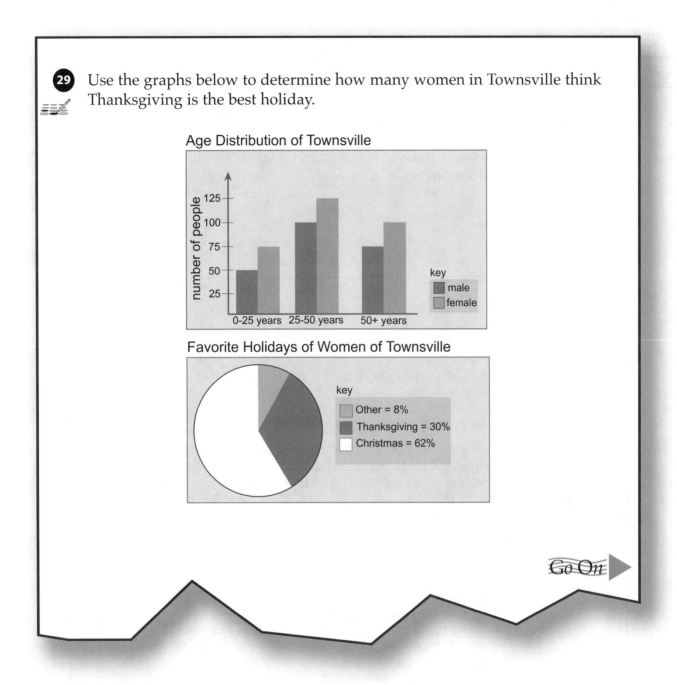

Age Distribution of Townsville

Favorite Holidays of Women of Townsville

key
- Other = 8%
- Thanksgiving = 30%
- Christmas = 62%

Go On ▶

Analysis: *The correct response is 90 women. To find the number of women in Townsville who like Thanksgiving best, you must first determine the total population of women in Townsville. This can be done from the graph of age distribution. From the graph, you can see that there are 75 women between 0 and 25 years old, 125 women between the ages of 25 and 50 years old, and 100 women 50 years or older. This is a total of 300 women. According to the pie chart key, 30% of women in Townsville like Thanksgiving best, so 30% of 300 will give the number of women who think Thanksgiving is the best holiday: 300* **x** *30% = 90 women.*

Question **30** *assesses*:

Strand E: Data Analysis and Probability

Standard 3: The student uses statistical methods to make inferences and valid arguments about real-world situations.

MA.E.3.4.2 Explains the limitations of using statistical techniques and data in making inferences and valid arguments.

Student Strategies:

There are two main types of statistics—descriptive statistics and inferential statistics. *Descriptive* statistics are concerned with collecting, summarizing, organizing, and presenting data. *Inferential* statistics are concerned with drawing conclusions about a population from evidence collected from some sample or sub-set of the population. Even when investigators make every effort to be fair and unbiased, errors can creep into a study because of some common mistakes. The most common mistakes include:

1. An assumption of cause and effect is made from correlational data. This is a kind of over-generalization. For example, just because your mother has coffee every morning and the sun comes up every morning, that doesn't mean your mother's coffee drinking is caused by the sunrise or that the sunrise is caused by your mother's coffee drinking.

2. The sample size might be too small.

3. The method for choosing the sample may be inappropriate. For example, the conclusions for many studies rely on a random sample. In a truly random sample, every member of the population has an equal chance of being chosen for the sample. Anything that gives part of the population a better chance of being selected than the rest, could introduce a hidden bias into the sample and threaten the conclusions that can be drawn from the data. Many "convenience samples" suffer from hidden bias. For example, conducting a survey on animal testing at a local veterinary clinic would be biased.

4. Sometimes an apparent result of an experiment or survey is be due to random variation. In other words, even though proper care is taken to choose an appropriate sample of the right size, the sample still does not accurately represent the entire population. This is because inferences from samples are based on probability and even highly improbable events do occasionally happen.

30 Quentin was conducting a survey of people's favorite movie types. On four separate nights, he conducted his survey at the four movie theaters in Baden. The results of his survey are shown in the table below. Which of the following can you determine from the information in the table?

Movie Theater

		Cinemania	Sticky Floors	Projections	Super Flicks
M o v i e T y p e s	Comedy	310	435	465	192
	Horror	242	273	271	151
	Romance	149	192	220	122
	Action	102	143	179	117

F. The most popular theater in Baden
G. The total population of Baden
H. The number of customers at Projections the night of the survey
I. The percent of the total votes "Comedy" received

Go On ▶

Analysis: *Choice I is correct. Because the total number of votes placed and the number of votes placed for "Comedy" is known, the percentage of votes received by "Comedy" can be found. Choice F is not correct because that would involve a separate survey about the most popular movie theater in Baden; you cannot assume the most popular movie theater is the movie theater where the most votes were placed because every customer in each theater may not have placed a vote. Also, most people attend a theater based on what movies are playing that night. They will only attend their favorite theater when it is showing the movie they want to see. Choice G is incorrect because you do not know if all the people in Baden went to a movie theater on the night of the survey. Choice H is incorrect for the same reason Choice F is incorrect—every customer may not have voted.*

This is the end of the Mathematics Practice Tutorial.
Until time is called, go back and check your work or answer
questions you did not complete. When you have finished, close
your workbook.

Name _____

Answer all the questions that appear in the Mathematics Practice Tutorial on this Answer Sheet.

1　Ⓐ　Ⓑ　Ⓒ　Ⓓ　　　　**2**　Ⓕ　Ⓖ　Ⓗ　Ⓘ　　　　**3**　Ⓐ　Ⓑ　Ⓒ　Ⓓ

4　Ⓕ　Ⓖ　Ⓗ　Ⓘ　　　　**5**　Ⓐ　Ⓑ　Ⓒ　Ⓓ　　　　**6**　Ⓕ　Ⓖ　Ⓗ　Ⓘ

7

8　Ⓐ　Ⓑ　Ⓒ　Ⓓ　　　　**9**　Ⓕ　Ⓖ　Ⓗ　Ⓘ

10　Ⓐ　Ⓑ　Ⓒ　Ⓓ　　　　**11**　Ⓕ　Ⓖ　Ⓗ　Ⓘ　　　　**12**

Fold and Tear Carefully Along Dotted Line.

13 Ⓐ Ⓑ Ⓒ Ⓓ   **14** Ⓕ Ⓖ Ⓗ Ⓘ   **15** Ⓐ Ⓑ Ⓒ Ⓓ

16 Ⓕ Ⓖ Ⓗ Ⓘ   **17**

	/	/	/	
•	•	•	•	•
0	0	0	0	0
1	1	1	1	1
2	2	2	2	2
3	3	3	3	3
4	4	4	4	4
5	5	5	5	5
6	6	6	6	6
7	7	7	7	7
8	8	8	8	8
9	9	9	9	9

18 Ⓐ Ⓑ Ⓒ Ⓓ

19 Ⓕ Ⓖ Ⓗ Ⓘ   **20** Ⓐ Ⓑ Ⓒ Ⓓ   **21**

	/	/	/	
•	•	•	•	•
0	0	0	0	0
1	1	1	1	1
2	2	2	2	2
3	3	3	3	3
4	4	4	4	4
5	5	5	5	5
6	6	6	6	6
7	7	7	7	7
8	8	8	8	8
9	9	9	9	9

Fold and Tear Carefully Along Dotted Line.

22 Ⓕ Ⓖ Ⓗ Ⓘ   **23** Ⓐ Ⓑ Ⓒ Ⓓ   **24** Ⓕ Ⓖ Ⓗ Ⓘ

25

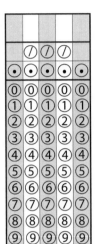

26 Ⓐ Ⓑ Ⓒ Ⓓ   **27** Ⓕ Ⓖ Ⓗ Ⓘ

28 Ⓐ Ⓑ Ⓒ Ⓓ   **29**

30 Ⓕ Ⓖ Ⓗ Ⓘ

Fold and Tear Carefully Along Dotted Line.

BLANK PAGE

Mathematics Assessment One

Directions for Taking the Mathematics Assessment One

On this section of the Florida Comprehensive Assessment Test (FCAT), you will answer 50 questions. For multiple-choice questions, you will be asked to pick the best answer out of four possible choices and fill in the answer in the answer bubble. On gridded-response questions, you will also fill your answer in answer bubbles, but you will fill in numbers and symbols corresponding to the solution you obtain for a question. Fill in the answer bubbles and gridded-response answer bubbles on the Answer Sheet on pages 155–159 to mark your selection.

Read each question carefully and answer it to the best of your ability. If you do not know an answer, you may skip the question and come back to it later.

Figures and diagrams with given lengths and/or dimensions are not drawn to scale. Angle measures should be assumed to be accurate. Use the formula sheet on page 33 and the conversion chart on page 34 to help you answer the questions. You will also be given a calculator to use.

When you finish, check your answers.

1 The table below shows the percent of votes each candidate received in a high school election for student-body president. The total number of votes cast was 1,680. If seniors cast 62.5% of the winning candidates votes, how many seniors voted for the winner?

Candidate	Percent of Vote
Todd Roe	15%
Libby Parker	5%
Steve Smith	60%
Terri Kirsch	20%

A. 630 votes
B. 700 votes
C. 1,008 votes
D. 1,050 votes

2 If *d* represents the number of doughnuts sold at a bakery, which of the following represents twice the number of doughnuts sold increased by 36?

F. $d + 38$
G. $(d + 2) + 36$
H. $2(d + 36)$
I. $2d + 36$

Go On ▶

3 Which of the following functions correctly expresses this written description of a function? The value of the independent variable, x, when increased by 2, will produce the corresponding value in the dependent variable.

A. $f(x) = 2$
B. $f(y) = 2$
C. $f(x) = 2 + y$
D. $f(x) = 2 + x$

4 Julio has a bag of 48 marbles. Of those marbles, one-third of them are red; the rest of the marbles are blue. If Julio draws two marbles out of the bag at the same time without looking, what is the probability that he will draw one marble of each color?

F. $\frac{3}{16}$

G. $\frac{2}{9}$

H. $\frac{1}{2}$

I. $\frac{2}{3}$

Go On

5 Fran is buying a rug to put in her bedroom. She wants a rug that will be 1 foot from the wall on each side. The dimensions of Fran's bedroom are shown in the diagram below. ESTIMATE the area of the rug Fran will need to buy to satisfy her conditions.

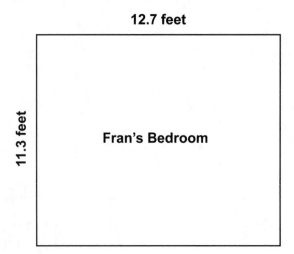

12.7 feet

11.3 feet

Fran's Bedroom

A. 156 square feet
B. 143 square feet
C. 99 square feet
D. 90 square feet

6 At a car factory, the workers can produce an average of 217.3 cars a day. ESTIMATE the number of cars they will produce in 7 months.

F. 1,400 cars
G. 4,000 cars
H. 28,000 cars
I. 40,000 cars

7 The table below shows the values of x and y for the equation $y - x^2 = 4$. What is the value of y when $x = -2$?

x	y
-3	13
-2	
-1	5
0	4
1	5
2	8
3	13

8 In a history class of 31 people, 16 people scored between 85% and 95% on the last test. Which of the following **must** be true?

A. The mean score will fall between 85% and 95%.
B. The mode of the test scores will fall between 85% and 95%.
C. The median of the test scores will fall between 85% and 95%.
D. The maximum score is closer to the mean score than the minimum score.

9 A local bakery uses an equation to determine the number of chocolate chips (n) to use for every dozen cookies (d). The equation is $244 - 2n = \frac{6}{d}$. Which of the following is equivalent to this equation?

F. $d = \frac{122 - n}{3}$

G. $d = \frac{3}{122 - n}$

H. $d = 238 - 2n$

I. $d = n - \frac{3}{n - 122}$

10 Given the diagram below of Circle P, which of the following is NOT true?

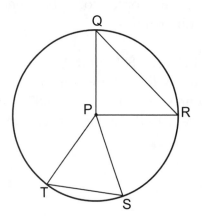

A. PR is congruent to PT.
B. Triangle STP is an equilateral triangle.
C. Angle RPQ + Angle SPR + Angle TPS + Angle QPT = 360°
D. Triangle RPQ is an isosceles triangle.

Go On ▶

11 Jim is required to pick a three-digit code to use as his password for his school email account. His password can contain any of the 26 letters in the alphabet in any location in the password, and he can use letters more than once. How many different combinations of letters does Jim have to choose from?

12 The Bounce-a-Ball Toy Company produces plastic balls for children to play with. The company produces spherical balls 2 feet in diameter. If the Bounce-a-Ball Toy Company wants to produce a ball with twice the diameter of its current ball, how much more volume will the new ball have compared to the original ball? Your answer will be in **cubic feet**. Round your answer to the nearest **tenth**.

13 Which of the following equations best fits the information contained on the scatterplot graph?

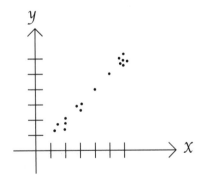

F. $x = y$
G. $2y = x$
H. $y = 2x$
I. $y = x^2$

Go On

14 Look at the picture of the paper clip below. Which of the answer choices could have resulted from a 180° rotation?

A.

C.

B.

D.

Go On

Copying is Prohibited

© **Englefield & Associates, Inc.**

15 Which of the following is equivalent to $2\sqrt{3}$ squared?

    F. $4\sqrt{3}$

    G. $4\sqrt{6}$

    H.   12

    I.   36

16 On the coordinate grid below, what is the distance between Point A and Point C? Round your answer to the nearest **hundredth**.

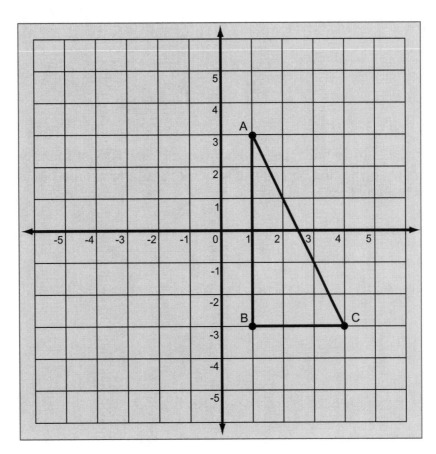

17 The chart below lists the final standings for the Moundsville Summer Baseball League. What was the **median** number of losses?

Team	Overall Record	
	Wins	Losses
Dragons	32	28
Hedgehogs	24	36
Llamas	33	27
Sloths	21	39
Wizards	40	20

A. 32
B. 30
C. 28
D. 27

18 Eloise planted 120 tulip bulbs last fall. Of those bulbs, 60% of them were eaten by squirrels before spring. If 12.5% of the remaining tulips bloomed in the spring, how many tulips bloomed in the spring?

F. 6 tulips
G. 9 tulips
H. 15 tulips
I. 48 tulips

19 Scott and Sarah are racing to a tree one mile away. Scott is able to run 23 miles per hour; Sarah is able to run 20 miles per hour. Because he is able to run faster, Scott moves his starting point back an extra tenth of a mile. If they both begin running at the same time, who will reach the tree first?

A. Scott will reach the tree first.
B. Sarah will reach the tree first.
C. They will reach the tree at the same time.
D. It is not possible to determine who will reach the tree first.

20 Which of the following is equivalent to eleven thousand nineteen and five hundredths?

F. 11,019.500
G. 11,019.05
H. 1,119.05
I. 11,190.05

21 Which of the following equations would be the correct representation for the data listed in the table of values shown below?

x	y
0	0
2	6
5	15
8	24

A. $y = x + 3$
B. $y = x + 6$
C. $y = 3x$
D. $y = 3 - x$

22 Jorge is a cook's assistant on a naval ship. One of his jobs is to peel potatoes. He can peel one potato in 12.3 seconds. How many **minutes** will it take him to peel 380 potatoes?

Go On

23 Each year, the seniors at John Glenn High School take a class trip. They are given several choices for possible destinations, and then each person gets to vote for his or her favorite. The circle graph below shows the vote distribution of the 476 seniors who voted. How many seniors voted to go to New Orleans?

24 Consider the following *x* values: 3, 6, 9 12, 15, and the corresponding *y* values: 1, 2, 3, 4, 5. Which of the following equations satisfies the values given?

F. $x = 2y$
G. $y = 2x$
H. $x = 3y$
I. $y = 3x$

Go On ▶

25 Which of the following sets of points could be used to create the graph of a line?

    A. (-4, -4); (0, 1); (5, 5)
    B. (-2, 1); (2, 1); (-2, -1)
    C. (3, -2); (-1, 0); (-9, 4)
    D. (4, 5); (5, 4); (6, 6)

26 A dartboard is mounted on a square piece of wood as shown below. What is the approximate total area of the parts of the square not covered by the dartboard?

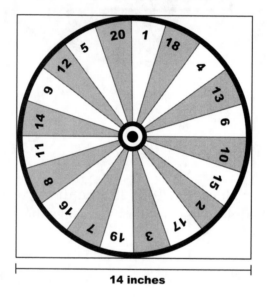

14 inches

    F. 42 square inches
    G. 56 square inches
    H. 154 square inches
    I. 196 square inches

Go On

27 Which of the following equations would be the correct representation for the data listed in the table of values shown below?

x	y
2	-1
4	1
8	5
10	7

A. $y = 2x$
B. $y = x + 4$
C. $y = x - 4$
D. $y = x - 3$

28 What is the slope of the line shown below?

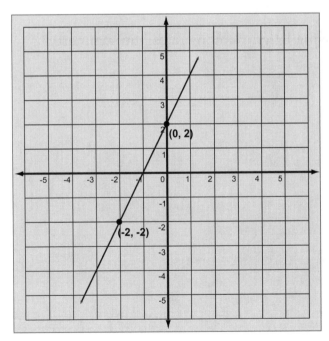

F. -4
G. -2
H. 2
I. 4

Go On

29 A random sampling of 200 people were asked to choose their favorite color. Of the people surveyed, 40% said they liked green best. In a population of 3,000 people, about how many people would like green best?

30 Marta goes to the mall to buy a new sweater. The sweater she chooses costs $36.99. When she goes to pay, she is informed that the sweater is marked down 20%. If sales tax is 6%, how much will Marta spend on the sweater?

A. $31.37
B. $30.04
C. $29.59
D. $27.37

31 Which of the following letters demonstrates line symmetry?

F. N
G. P
H. T
I. S

32 Which of the following best illustrates the basic graph of $y = x + 3$ as the quadrant I portion would appear?

A.

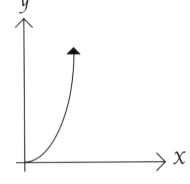

C.

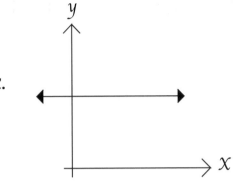

B.

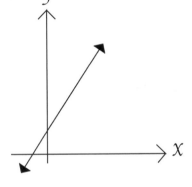

D.

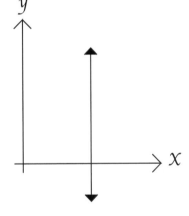

Go On

33 On any given day at Jan's Seaside Snack Bar, hamburgers (x) outsell hotdogs (y) by a two to one ratio. It has been this way for years, and Jan always places her wholesale food orders based on this simple relationship. Which of the following describe a way in which this relationship could be shown as a linear equation?

F. $x^2 = y$
G. $x = 2y$
H. $x = y^2$
I. $2x = y$

34 What number is missing from the pattern below?

$$63, 64, 8, \underline{\quad}, 3, 4, 2$$

35 Which of the following expressions will result in the greatest number when multiplied by -1?

A. $1 + (-2)^2$
B. $1 - 2^2$
C. $1 + (-1)^3$
D. $1 - 1^3$

Go On

36 Manolo needs to call his friend June but cannot remember her phone number. All he can remember is that the first number is a 7. If June's phone number is a standard seven-digit phone number, what are the odds Manolo will be able to guess her phone number on the first try?

F. $\frac{1}{7}$

G. $\frac{1}{10}$

H. $\frac{1}{1,000,000}$

I. $\frac{1}{10,000,000}$

37 Which of the following CANNOT be assumed about the parallelogram shown below?

A. The sum of the interior angles of ABCD is 360°.
B. AB ∥ DC
C. AB + AD = DC + BC
D. BC ⊥ CD

Go On ▶

38 A clothing company is designing a symbol to use as its logo. The logo is made up of a square, two equilateral triangles, and two semicircles. In the diagram of the logo shown below, you are given the length of the diagonal of the square and the height of one of the triangles. What is the area of the logo? Round your answer to the nearest **tenth**.

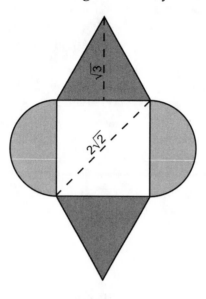

39 Hortense is increasing the size of her garden. The original dimensions of the garden are 20 yards by 25 yards. If the garden is lengthened by 10 yards on each side, how will this affect the perimeter of the garden?

F. The perimeter will increase by 80 yards.
G. The perimeter will increase by 40 yards.
H. The perimeter will increase by 20 yards.
I. The perimeter will increase by 10 yards.

40 Which of the following represents the **greatest** number?

    A. $(0.25)^3$

    B. $(2)^{-5}$

    C. $(-1)^5$

    D. $(3)^{-2}$

41 If triangle ABC were translated such that point B's new coordinates were (6, 0), what would be the new coordinates for point C?

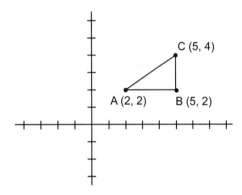

    F. (5, 0)

    G. (6, 1)

    H. (6, 2)

    I. (7, 2)

42 Chip and Jen are designing a birdhouse for a school project. For the roof and base of the birdhouse, they used pieces of wood measuring 6 inches by 6 inches by 0.5 inch. For the four sides of the birdhouse, they used pieces of wood measuring 8 inches by 6 inches by 0.5 inch. After they had assembled these pieces into a rectangular prism, they added a small piece of wood measuring 1 inch by 3 inches by 0.5 inch to serve as a ledge. To complete the project, they cut a cylinder with a diameter of 2 inches out of the front of the birdhouse for the birds to enter. What is the total volume of wood in the completed birdhouse? Your answer should be in **cubic inches** rounded to the nearest **tenth**.

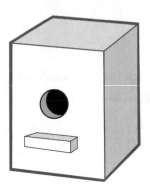

43 A common garden snail travels at a speed of 0.03 miles per hour. The giant tortoise travels at a speed of 0.17 miles per hour. Both the snail and the tortoise took the same amount of time to travel from their respective starting points to the tree. If the tortoise and the snail both arrive at a tree at the exact same moment, how far away from the tree was the garden snail when the tortoise was 2 miles away from the tree? Round to the nearest **hundredth**.

44 If pentagon ABCDE is reflected across the *y*-axis, what are the new coordinates of point D?

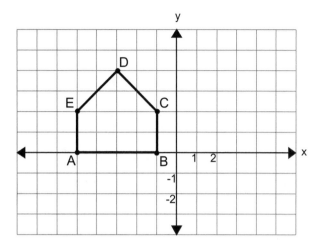

A. (3, 4)
B. (-3, -4)
C. (1, 0)
D. (-1, 0)

45 Using the Pythagorean theorem, what is the distance between C and A?

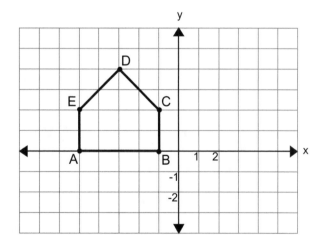

F. 20
G. $2\sqrt{5}$
H. 5
I. $\sqrt{6}$

Go On ▶

46 Which of the following statements best describes the combined area of triangles ABC and AZY?

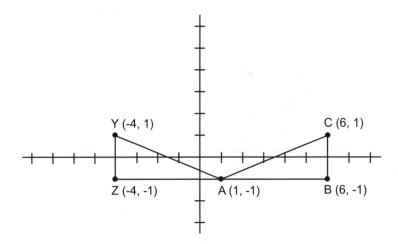

A. The areas of the triangles cannot be determined, since the hypotenuse of both triangles has a portion located within the negative quadrants III and IV.

B. Since these right triangles are congruent, they could combine to form a rectangle, and its area would then be found by multiplying length times width.

C. Since none of the lengths are known, the areas cannot be found.

D. The areas of the triangles cannot be found, since the base of each triangle is located below the *x*-axis, and those values are negative.

 Go On ▶

47 Which of the following pictures best demonstrates the basic premise of the Pythagorean theorem, using area to demonstrate the reason the theorem works?

F.

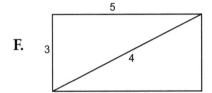

H.

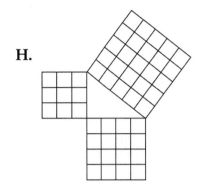

G.

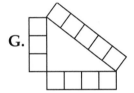

I.

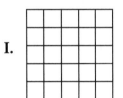

Go On ▶

48 Dean wants to know how tall a tree in his yard is, but he does not have the proper tools to measure it by hand. He does know that he is 6 feet tall, and he is able to measure the length of his shadow (12 feet) and the length of the shadow of the tree (36 feet). How tall is the tree in **inches**?

49 Drive-Away-Today auto dealership sells a car for $1,400.00 dollars above its wholesale cost. The manufacturer will also give the dealership a $200.00 incentive fee for each car that is sold. Which of the following functions describes the total profit, $f(c)$, for number of cars sold, c?

A. $f(c) = \$1,400c + \$200$
B. $f(c) = \$1,400 + \$200c$
C. $f(c) = \$1,600c$
D. $f(c) = \$1,400c$

50 Brenda received $120,000.00 when a relative's estate was settled. She wants to invest this money. She plans to follow the advice of a financial advisor who has told her to invest 50% of this money in a savings account, 20% in bonds, and the remaining amount in stocks. How much money will be invested in stocks?

F. $20,000.00

G. $28,000.00

H. $36,000.00

I. $40,000.00

This is the end of Mathematics Assessment One.
Until time is called, go back and check your work or answer
questions you did not complete. When you have finished, close
your workbook.

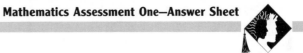

Name _____

Answer all the questions that appear in the Mathematics Assessment One on this Answer Sheet.

1 Ⓐ Ⓑ Ⓒ Ⓓ   **2** Ⓕ Ⓖ Ⓗ Ⓘ   **3** Ⓐ Ⓑ Ⓒ Ⓓ

4 Ⓕ Ⓖ Ⓗ Ⓘ   **5** Ⓐ Ⓑ Ⓒ Ⓓ   **6** Ⓕ Ⓖ Ⓗ Ⓘ

7   **8** Ⓐ Ⓑ Ⓒ Ⓓ   **9** Ⓕ Ⓖ Ⓗ Ⓘ

10 Ⓐ Ⓑ Ⓒ Ⓓ   **11**   **12**

Go On ▶

13 Ⓕ Ⓖ Ⓗ Ⓘ   **14** Ⓐ Ⓑ Ⓒ Ⓓ   **15** Ⓕ Ⓖ Ⓗ Ⓘ

16

17 Ⓐ Ⓑ Ⓒ Ⓓ   **18** Ⓕ Ⓖ Ⓗ Ⓘ

19 Ⓐ Ⓑ Ⓒ Ⓓ   **20** Ⓕ Ⓖ Ⓗ Ⓘ   **21** Ⓐ Ⓑ Ⓒ Ⓓ

22   **23**   **24** Ⓕ Ⓖ Ⓗ Ⓘ

Fold and Tear Carefully Along Dotted Line.

Go On ▶

Fold and Tear Carefully Along Dotted Line.

25 Ⓐ Ⓑ Ⓒ Ⓓ   **26** Ⓕ Ⓖ Ⓗ Ⓘ   **27** Ⓐ Ⓑ Ⓒ Ⓓ

28 Ⓕ Ⓖ Ⓗ Ⓘ   **29**   **30** Ⓐ Ⓑ Ⓒ Ⓓ

31 Ⓕ Ⓖ Ⓗ Ⓘ   **32** Ⓐ Ⓑ Ⓒ Ⓓ   **33** Ⓕ Ⓖ Ⓗ Ⓘ

34    **35** Ⓐ Ⓑ Ⓒ Ⓓ   **36** Ⓕ Ⓖ Ⓗ Ⓘ

Go On ▶

37   Ⓐ   Ⓑ   Ⓒ   Ⓓ

38
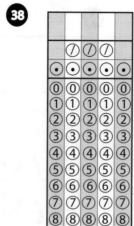

39   Ⓕ   Ⓖ   Ⓗ   Ⓘ

40   Ⓐ   Ⓑ   Ⓒ   Ⓓ

41   Ⓕ   Ⓖ   Ⓗ   Ⓘ

42
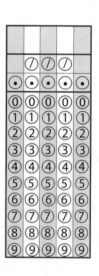

43
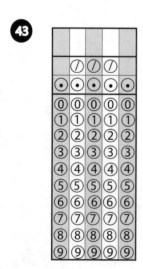

44   Ⓐ   Ⓑ   Ⓒ   Ⓓ

45   Ⓕ   Ⓖ   Ⓗ   Ⓘ

Fold and Tear Carefully Along Dotted Line

Go On ▶

46 Ⓐ Ⓑ Ⓒ Ⓓ　　　　　**47** Ⓕ Ⓖ Ⓗ Ⓘ　　　　**48**

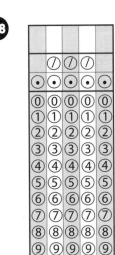

49 Ⓐ Ⓑ Ⓒ Ⓓ　　　　　**50** Ⓕ Ⓖ Ⓗ Ⓘ

Fold and Tear Carefully Along Dotted Line.

BLANK PAGE

Mathematics Assessment One: Skills Chart

Question	Standard	Answer	Keywords
1	MA.E.3.4.1	A	Performs real-world statistical experiments
2	MA.A.1.4.3	I	Understands numbers in real-world situations
3	MA.D.1.4.1	D	Interprets data
4	MA.E.2.4.2	G	Determines probability for simple and compound events as well as independent/dependent events
5	MA.B.3.4.1	C	Solves real-world problems involving estimates
6	MA.A.4.4.1	I	Uses estimation strategies to predict results
7	MA.D.1.4.1	8	Analyzes functions using variables
8	MA.E.1.4.2	C	Calculates measures of central tendency (mean, median, and mode)
9	MA.D.1.4.1	G	Interprets data
10	MA.C.1.4.1	B	Uses properties of geometric shapes
11	MA.E.2.4.1	17,576	Determines probabilities
12	MA.D.1.4.2	29.3	Determines impact when changing parameters of given functions
13	MA.D.2.4.1	F	Uses systems of equations to solve problems
14	MA.C.2.4.1	C	Understands geometric concepts
15	MA.A.3.4.1	H	Understands square roots
16	MA.C.3.4.1	6.71	Applies geometric properties to solve problems
17	MA.E.1.4.2	C	Calculates measures of central tendency (mean, median, and mode)
18	MA.A.2.4.2	F	Understands and uses the real number system
19	MA.B.2.4.2	A	Solves real-world problems involving measures
20	MA.A.1.4.1	G	Associates written word names with real numbers
21	MA.D.1.4.1	C	Analyzes functions using variables
22	MA.B.2.4.1	77.9	Solves real-world problems involving rated measures
23	MA.E.1.4.1	119	Interprets data
24	MA.D.1.4.1	H	Analyzes functions using variables
25	MA.C.3.4.2	C	Using a graph applies properties of two- and three-dimensional figures

Mathematics Assessment One: Skills Chart

Question	Standard	Answer	Keywords
26	MA.B.1.4.1	F	Uses graphic models to derive formulas for finding perimeter
27	MA.D.1.4.1	D	Describes functions using tables
28	MA.C.3.4.2	H	Using a graph applies properties of two- and three-dimensional figures
29	MA.E.1.4.3	1200	Analyzes real-world data to make predictions
30	MA.A.3.4.3	A	Adds, subtracts, multiplies, and adds real numbers
31	MA.C.2.4.1	H	Understands geometric concepts
32	MA.D.1.4.1	B	Analyzes functions using graphs
33	MA.D.2.4.1	G	Represents real-world problem situations using equations
34	MA.D.1.4.1	9	Analyzes patterns
35	MA.A.3.4.1	B	Understands operations
36	MA.E.2.4.2	H	Determines probabilities
37	MA.C.1.4.1	D	Uses properties of geometric shapes to construct formal and informal proofs
38	MA.B.1.4.1	10.6	Use graphic models to find area
39	MA.D.1.4.2	G	Calculate measures and dispersion
40	MA.A.1.4.2	D	Understands different ways numbers are represented in the real-world
41	MA.C.2.4.1	H	Understands geometric concepts
42	MA.B.1.4.1	131.9	Uses measures to solve problems
43	MA.B.2.4.2	0.35	Solves real-world problems involving rated measures
44	MA.C.2.4.1	A	Understands geometric concepts
45	MA.C.3.4.1	G	Represents and applies geometric properties to solve problems
46	MA.C.3.4.2	B	Using a graph applies properties of two- and three-dimensional figures
47	MA.C.1.4.1	H	Uses properties of geometric shapes to construct formal and informal proofs
48	MA.B.1.4.3	216	Relates the concepts of measurement to similarity and proportionality in real-world situations
49	MA.D.1.4.1	C	Analyzes functions
50	MA.A.1.4.4	H	Understands that numbers can be represented in different ways

Mathematics Assessment One: Answer Key

1 The table below shows the percent of votes each candidate received in a high school election for student-body president. The total number of votes cast was 1,680. If seniors cast 62.5% of the winning candidates votes, how many seniors voted for the winner?

Candidate	Percent of Vote
Todd Roe	15%
Libby Parker	5%
Steve Smith	60%
Terri Kirsch	20%

Analysis: *Choice A is correct.* Steve Smith received the most votes with 60% of the total, or 0.60 x 1,680 = 1,008 votes. Of the votes cast for Steve Smith, seniors cast 62.5%, so 1,008 x 0.625 = 630 seniors voted for the winner of the election, Steve Smith. Choice B is just a guess. Choice C is incorrect because 1,008 is the total number of votes cast for the winner, not the number of seniors that voted for the winner. Choice D is incorrect because 1,050 is 62.5% of the total votes, not the votes cast for the winner.

2 If d represents the number of doughnuts sold at a bakery, which of the following represents twice the number of doughnuts sold increased by 36?
Analysis: *Choice I is correct.* The expression that represents twice the number of doughnuts sold increased by 36 is $2d + 36$. The expression in choice F represents the number of doughnuts sold increased by 36 + 2, not twice the number of doughnuts sold increased by 36. The expression in choice G represents the number of doughnuts sold plus 2, then increased by 36. The expression in choice H represents the number of doughnuts increased by 36, then multiplied by 2.

3 Which of the following functions correctly expresses this written description of a function? The value of the independent variable, x, when increased by 2, will produce the corresponding value in the dependent variable.
Analysis: *Choice D is correct.* The value of $f(x)$ is found by increasing x by 2, which means 2 is added to x. Choice A is incorrect because $f(x)$ is always 2 no matter what the value of x is. Choice B is incorrect for the same reason as Choice A and also because the independent variable is supposed to be x, not y. Choice C is incorrect because the independent variable is supposed to be x, not y.

4 Julio has a bag of 48 marbles. Of those marbles, one-third of them are red; the rest of the marbles are blue. If Julio draws two marbles out of the bag at the same time without looking, what is the probability that he will draw one marble of each color?
Analysis: *Choice G is correct.* The probability that Julio will draw one marble of each color is 2/9: $1/3 \times 2/3 = 2/9$. Choice F appears to be a guess and is incorrect. Choice H is incorrect because 1/2 is the probability of drawing a certain color marble out of the bag if there were an equal number of each color of marble in the bag. Choice I is incorrect because 2/3 is the probability of drawing a blue marble out of the bag if one marble is drawn.

5 Fran is buying a rug to put in her bedroom. She wants a rug that will be 1 foot from the wall on each side. The dimensions of Fran's bedroom are shown in the diagram below. ESTIMATE the area of the rug Fran will need to buy to satisfy her conditions.

12.7 feet

11.3 feet

1ft
10.7 ft
1ft   9.3 ft   Rug Area   1ft
1ft

Analysis: *Choice C is correct.* First round off the room's dimensions to the nearest foot. This makes the room approximately 11 ft by 13 ft. Since she wants the rug to be 1 foot from the wall on each side, subtract 2 feet from each of the room's dimensions. Thus the rugs dimensions are 9 ft x 11 ft = 99 sq ft. Choices A and B are incorrect because 2 ft should be subtracted from both the length and width of the room before finding the area of the rug. In addition, in Choice A, the width of the room is also rounded up instead of properly being rounded down. Choice D is incorrect because the 10.7 ft is rounded down to 10 instead of rounded up to 11.

Mathematics Assessment One: Answer Key

6 At a car factory, the workers can produce an average of 217.3 cars a day. ESTIMATE the number of cars they will produce in 7 months.
Analysis: *Choice I is correct.* At the rate given, the factory will produce approximately 200 cars each day for 30 days each month for 7 months: 200 x 30 x 7 = 42,000 cars, which is about 40,000 cars. Choice F is the number of cars the factory will produce in 7 days: 200 cars x 7 days = 1,400 cars. Choice G is the number of cars the factory will produce in 20 days: 200 cars x 20 days = 4,000 cars. Choice H is the number of cars the factory will produce in 7 months of production for 20 days per month: 200 cars x 20 days x 7 months = 28,000 cars.

7 The table below shows the values of x and y for the equation $y - x^2 = 4$. What is the value of y when $x = -2$?

x	y
-3	13
-2	
-1	5
0	4
1	5
2	8
3	13

Analysis: *The correct answer is 8.* To solve this problem, plug -2 into the equation for x, and solve for y: $y - (-2^2) = 4$; $y - 4 = 4$; $y = 4 + 4$; $y = 8$.

8 In a history class of 31 people, 16 people scored between 85% and 95% on the last test. Which of the following **must** be true?
Analysis: *Choice C is correct.* On this test more than half of the class scored between 85% and 95%. The median is the middle number when the scores are placed in order from least to greatest, so the middle number must fall somewhere between 85% and 95%. The mean is the arithmetic average of the scores. With nearly half of the scores falling outside the 85% to 95% range, it is possible for the average score to also be outside this range of scores. Choice A is incorrect. The mode of the test scores is the number that appears most often. It is also possible that the mode of the scores will be outside the 85% and 95% range, so Choice B is incorrect. It is impossible to know where the maximum or the minimum scores fall by knowing only that just over half of the scores are in the 85% to 95% range, so Choice D is incorrect.

9 A local bakery uses an equation to determine the number of chocolate chips (n) to use for every dozen cookies (d). The equation is $244 - 2n = 6/d$. Which of the following is equivalent to this equation?
Analysis: *Choice G is correct.* First multiply both sides of the equation by d to get $d(244 - 2n) = 6$. Next divide both sides of the equation by the items in parentheses to isolate d on the left side of the equation: $d = 6/(244 - 2n)$. This reduces to $d = 3/122 - n$ because 6, 244, and -2 are all divisible by 2. Choice F is incorrect because the numerator and denominator are inverted. Choice H is incorrect because 6 was subtracted from 244 rather than dividing the numerator and denominator by 2. Choice I is incorrect because 122 is subtracted from n instead of n being subtracted from 122.

10 Given the diagram below of Circle P, which of the following is NOT true?

Analysis: *Choice B is correct.* We cannot say that triangle STP is an equilateral triangle because we do not know whether the length of ST is equal to the lengths of PS and PT. This may be true or may be false. However, stating definitely that triangle STP is an equilateral triangle when it may not be, is clearly false. Choice A is incorrect because PR is congruent to PT, since all radii have equal length. Choice C is incorrect because Angle RPQ + Angle SPR + Angle TPS + Angle QPT = 360°, since the sum of the interior angles of a circle = 360°. Choice D is incorrect because triangle RPQ is an isosceles triangle since two of its sides are radii which, of course, are of equal length.

Mathematics Assessment One: Answer Key

11 Jim is required to pick a three-digit code to use as his password for his school email account. His password can contain any of the 26 letters in the alphabet in any location in the password, and he can use letters more than once. How many different combinations of letters does Jim have to choose from?

Analysis: *The correct answer is 17,576.* To solve this problem find 26^3: 26 x 26 x 26 = 17,576.

12 The Bounce-a-Ball Toy Company produces plastic balls for children to play with. The company produces spherical balls 2 feet in diameter. If the Bounce-a-Ball Toy Company wants to produce a ball with twice the diameter of its current ball, how much more volume will the new ball have compared to the original ball? Your answer will be in **cubic feet**. Round your answer to the nearest **tenth**.

Analysis: *The correct answer is 29.3 ft³.* The formula for the volume of a sphere is $V = \frac{4}{3}\pi r^3$. Since the original ball has a diameter of 2 feet, it has a radius of 1 foot. Calculating its volume using the formula: $V = \frac{4}{3}\pi r^3$; $V = \frac{4}{3}\pi 1^3$;

$V = \frac{4}{3}\pi$; $V \approx 4.2$ ft³. Since the new ball has a diameter twice as large as the original, its radius must also be twice as large, or 2 feet. Calculating its volume using the formula:

$V = \frac{4}{3}\pi r^3$; $V = \frac{4}{3}\pi 2^3$; $V = \frac{4}{3}\pi (8)$;

$V \approx 33.5$ ft³. Therefore, the new ball's volume is 29.3 cubic feet larger than that of the original ball

(33.5 − 4.2 = 29.3).

13 Which of the following equations best fits the information contained on the scatterplot graph?

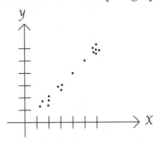

Analysis: *Choice F is correct.* The points seem to roughly follow a $x = y$ line. Choice G is incorrect because $2y = x$ is the same as $y = 1/2x$ which has a slope half as steep as $x = y$. Choice H is incorrect because $y = 2x$ has a slope twice as steep as $x = y$. Choice I is incorrect because the graph of $y = x^2$ is a parabola.

14 Look at the picture of the paper clip below. Which of the answer choices could have resulted from a 180° rotation?

Analysis: *Choice C is correct.*

Choice A is incorrect because this picture represents a reflection across a horizontal line, not a rotation. Choice B is incorrect because this picture represents a reflection across a vertical line, not a rotation. Choice D is incorrect because this picture represents a rotation of 0° or 360°.

15 Which of the following is equivalent to $2\sqrt{3}$ squared?

Analysis: *Choice H is correct.* $2\sqrt{3}$ squared = $(2\sqrt{3})^2 = 2^2 \times \sqrt{3}^2 = 2^2 \times \sqrt{9} = 4 \times 3 = 12$. Choice F is incorrect because $\sqrt{3}$ squared is equivalent to $\sqrt{3} \times \sqrt{3} = 3$, not $\sqrt{3}$. Choice G is incorrect because $\sqrt{3}$ squared is equivalent to 3, not $\sqrt{6}$. Choice I is incorrect because $\sqrt{3}$ squared is equivalent to 3, not 3^2, or 9.

16 On the coordinate grid below, what is the distance between Point A and Point C? Round your answer to the nearest **hundredth**.

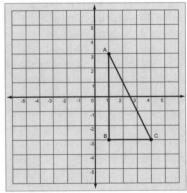

Analysis: *The correct answer is 6.71.* The triangle is a right triangle, so the distance between Point A and Point C can be found by using the Pythagorean theorem. Note that the length of segment BC is 3 and the length of segment AB is 6. Also remember that segment AC, the longest side or hypotenuse, must be the lone variable, c. $6^2 + 3^2 = c^2$; $36 + 9 = c^2$; $45 = c^2$; $c = \sqrt{45}$. Therefore, the distance between Point A and Point C equals $\sqrt{45}$, or 6.71.

Copying is Prohibited

Mathematics Assessment One: Answer Key

17 The chart below lists the final standings for the Moundsville Summer Baseball League. What was the **median** number of losses?

Team	Overall Record	
	Wins	**Losses**
Dragons	32	28
Hedgehogs	24	36
Llamas	33	27
Sloths	21	39
Wizards	40	20

Analysis: *Choice C is correct.* The median number in a set of data is the middle number when the items are placed in order from least to greatest. Placing the numbers in the loss column in increasing numerical order yields: 20, 27, 28, 36, 39. The middle number is 28, so it is the median. Choice A is incorrect because 32 is the median of the numbers in the Win column. Choice B is incorrect because 30 is the median of the numbers in both the Win and Loss columns combined. Choice D is incorrect because 27 is the number of losses recorded by the Llamas and the third item in the Loss column. It is not the median or middle number of the set because the numbers in the Loss column are not yet placed in proper order from least to greatest.

18 Eloise planted 120 tulip bulbs last fall. Of those bulbs, 60% of them were eaten by squirrels before spring. If 12.5% of the remaining tulips bloomed in the spring, how many tulips bloomed in the spring?

Analysis: *Choice F is correct.* If 60% of the 120 tulip bulbs Eloise planted last fall were eaten by squirrels, 40%, or 120 x 0.40 = 48 tulip bulbs, remained. Of those remaining 48 bulbs, 12.5% bloomed in the spring, so 0.125 x 48 = 6 bulbs bloomed. Choice G is incorrect because 9 is the number of tulips that may have bloomed if 60% of the bulbs remained rather than 40%. Choice H is incorrect because 15 tulips are 12.5% of the original 120 bulbs. To solve this problem you must first find the number of bulbs left after the squirrels ate 60% of the bulbs, then find the number of bulbs that bloomed from that number. Choice I is incorrect because 48 is the number of bulbs that remained after the squirrels ate 60%.

19 Scott and Sarah are racing to a tree one mile away. Scott is able to run 23 miles per hour; Sarah is able to run 20 miles per hour. Because he is able to run faster, Scott moves his starting point back an extra tenth of a mile. If they both begin running at the same time, who will reach the tree first?

Analysis: *Choice A is correct.* To find how long it will take each runner to run one mile, divide the runner's rate of speed by 60 minutes. Scott runs one mile in 2.6 minutes: 60 minutes per hour ÷ 23 miles per hour = 2.6 minutes per mile. To find out how long it will take Scot to run 1.1 miles, multiply the time it takes him to run one mile by 1.1 miles: 2.6 minutes per mile x 1.1 miles = 2.86 minutes. Sarah will reach the tree in three minutes: 60 minutes per hour ÷ 20 miles per hour = 3 minutes per mile. Scott will reach the tree first.

20 Which of the following is equivalent to eleven thousand nineteen and five hundredths?

Analysis: *Choice G is correct.* Eleven thousand nineteen and five hundredths is equivalent to 11,019.05. When properly used, the word "and" indicates where the decimal point goes. The first place to the right of the decimal is the tenths place, the second place to the right of the decimal is the hundredths place, and the third place to the right of the decimal is the thousandths place. Choice F is incorrect because 11,019.500 is equivalent to eleven thousand nineteen and five tenths (or equally, five hundred thousandths), not eleven thousand nineteen and five hundredths. Choice H is incorrect because 1,119.05 is equivalent to one thousand one hundred nineteen and five hundredths, not eleven thousand nineteen and five hundredths. Choice I is incorrect because 11,190.05 is equivalent to eleven thousand one hundred ninety and five hundredths, not eleven thousand nineteen and five hundredths.

21 Which of the following equations would be the correct representation for the data listed in the table of values shown below?

x	y
0	0
2	6
5	15
8	24

Analysis: *Choice C is correct.* If you're not sure how to solve a problem like this, plug each x into the equation and see if it correctly calculates the y value in the table. All of the x values substituted into the equation must yield the same y values as in the table for that choice to be correct. Choices A, B, and D all fail on the very first value of $x = 0$. None of these choices yield $y = 0$ when $x = 0$ is plugged into the equation.

Mathematics Assessment One: Answer Key

22 Jorge is a cook's assistant on a naval ship. One of his jobs is to peel potatoes. He can peel one potato in 12.3 seconds. How many minutes will it take him to peel 380 potatoes?
Analysis: *The correct answer is 77.9 minutes.* If you multiply the speed at which Jorge can peel one potato by the number of potatoes he needs to peel: 12.3 seconds x 380 potatoes = 4,674 seconds. Convert this answer from seconds to minutes to complete the problem: 4,674 ÷ 60 = 77.9 minutes. It will take Jorge 77.9 minutes to peel 380 potatoes.

23 Each year, the seniors at John Glenn High School take a class trip. They are given several choices for possible destinations, and then each person gets to vote for his or her favorite. The circle graph below shows the vote distribution of the 476 seniors who voted. How many seniors voted to go to New Orleans?

Analysis: *The correct answer is 119.* Students who voted for New Orleans forms a 90° angle on the circle graph. That is equivalent to 25% of the graph, so it equals 25% of the votes. To find how many seniors voted to go to New Orleans, multiply the total votes by 25%: 476 x 0.25 = 119.

24 Consider the following x values: 3, 6, 9 12, 15, and the corresponding y values: 1, 2, 3, 4, 5. Which of the following equations satisfies the values given?
Analysis: *Choice H is correct.* Three times the y values given will yield the corresponding x value.

X	3	6	9	12	15
Y	1	2	3	4	5

25 Which of the following sets of points could be used to create the graph of a line?
Analysis: *Choice C is correct.* In order for a set of points to create the graph of a line, they have to be collinear. That is, all the members of the set need to exist on the same straight line. If you carefully plot all of the points in each of these sets, you will find that only the members of the set in Choice C all exist on the same straight line. All of the other sets form triangles.

26 A dartboard is mounted on a square piece of wood as shown below. What is the approximate total area of the parts of the square not covered by the dartboard?

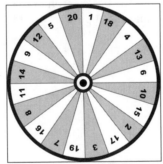

Analysis: *Choice F is correct.* Notice that the diameter of the circle equals the length of the square's side, 14 inches. Since the circle's radius is half its diameter, use $r = 7$ inches the formula for the area of the circle, $A = \pi r^2$: $A = \pi 7^2$; 3.1415 x 49 ≈ 154.1, or 154 square inches. The area of the square is found by multiplying the length of the sides: 14 inches x 14 inches = 196 square inches. Subtract the area of the dartboard from the area of the square to find the area not covered by the dartboard: 196 square inches – 154 square inches ≈ 42 sq in. Choice G is incorrect because it is the perimeter of the entire wooden square. Choice H is incorrect because it is the entire area of the circular dartboard. Choice I is incorrect because it is the area of the wooden square.

27 Which of the following equations would be the correct representation for the data listed in the table of values shown below?

x	y
2	-1
4	1
8	5
10	7

Analysis: *Choice D is correct.* Since the y values are all 3 less than the x values, a correct equation for this table would be $y = x - 3$.

28 What is the slope of the line shown below?

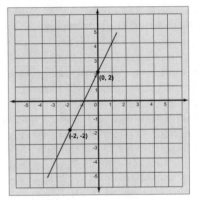

Analysis: *Choice H is correct.* The slope of a line is: (its change in y)/(its change in x). An easy way to remember this is: slope = $\frac{rise}{run}$. The line rises from left to right, so the slope is positive. Look at two points on the line. In order to get from point (-2, -2) to point (0, 2), you must go up 4 and right 2. That is, while the y value increases by four units, the x value moves right two units. This is a slope of 4/2 which reduces to 2/1 or 2.

Mathematics Assessment One: Answer Key

29 A random sampling of 200 people were asked to choose their favorite color. Of the people surveyed, 40% said they liked green best. In a population of 3,000 people, about how many people would like green best?
Analysis: *The correct answer is 1200.* If the sampling is representative of the population and 40% of the sample chose green as their favorite color, it can be concluded that 40% of the overall population of 3,000 people would probably select green as well. To find 40% of 3,000, multiply 3,000 people x 0.40 = 1,200 people.

30 Marta goes to the mall to buy a new sweater. The sweater she chooses costs $36.99. When she goes to pay, she is informed that the sweater is marked down 20%. If sales tax is 6%, how much will Marta spend on the sweater?
Analysis: *Choice A is correct.* From an original cost of $36.99 the sweater is marked down 20%, so the sale price of the sweater is 80% of the original cost: $36.99 x 0.80 = $29.59. Next find the tax by multiplying the sale price by 0.06: $1.78. Finally add the tax to the sale price of the sweater to find how much will Marta spend on the sweater: $29.59 + $1.78 = $31.37. Choice B appears to be a guess and is incorrect. Choice C is incorrect because $29.59 is the cost of the sweater before the tax is added. Choice D is incorrect because $27.37 is the original cost of the sweater reduced by 26%, rather than reduced by 20% with 6% tax added to that amount.

31 Which of the following letters has line symmetry?
Analysis: *Choice H is correct.* An item has line symmetry if it can be divided in half along a line, with the two halves being mirror images of each other. Of the letters N, P, T, and S, only the T can be divided in half with the two halves being mirror images of each other. The letter S and N both have what is called radial or rotational symmetry. When rotated 180° (turned upside down) and they look exactly the same. The letter P has no symmetry. Some letters have line symmetry, some have rotational symmetry, some have both, and some have neither.

32 Which of the following best illustrates the basic graph of $y = x + 3$ as the quadrant I portion would appear?
Analysis: *Choice B is correct.*

This is a linear equation with a positive slope and a y-intercept of 3. Choice A represents a curved exponential graph, while Choice C is a horizontal line with a slope of 0. Choice D is a vertical line with an undefined slope.

33 On any given day at Jan's Seaside Snack Bar, hamburgers (x) outsell hotdogs (y) by a two to one ratio. It has been this way for years, and Jan always places her wholesale food orders based on this simple relationship. Which of the following describe a way in which this relationship could be shown as a linear equation?
Analysis: *Choice G is correct.* Set the number of hotdogs sold as y, and the number of hamburgers sold as x. Therefore, $x = 2y$. Values for each can then be easily determined and plotted as a line.

34 What number is missing from the pattern below?
    63, 64, 8, __, 3, 4, 2
Analysis: *The correct answer is 9.* The pattern can be described as +1, then the square root of the result, then + 1, then the square root of the result, etc. Since 8 is the square root of 64, the next number would be 9: 8 + 1 = 9.

35 Which of the following expressions will result in the greatest number when multiplied by -1?
Analysis: *Choice B is correct.* Calculate to find which of the expressions will result in the greatest number when multiplied by -1. For Choice B, $-1(1 - 2^2) = -1(1 - 4) = -1(-3) = 3$. For choice A, $-1(1 + (-2)^2) = -1(1 + 4) = -1(5) = -5$. For Choice C; $-1(1 + (-1)^3) = -1(1 - 1) = -1(0) = 0$. For Choice D, $-1(1 - 1^3) = -1(1 - 1) = -1(0) = 0$.

Mathematics Assessment One: Answer Key

36 Manolo needs to call his friend June but cannot remember her phone number. All he can remember is that the first number is a 7. If June's phone number is a standard seven-digit phone number, what are the odds Manolo will be able to guess her phone number on the first try?
Analysis: *Choice H is correct.* June's phone number is a standard seven-digit phone number, but Manolo can remember the first number, so he only has to guess the other 6. The chance that Manolo will be able to guess her phone number on the first try is $1/10^6$, or $1/1,000,000$. This is because there are 10 possible numbers for each of the six digits he can't remember, so the chance that he will dial any digit correctly is $1/10$ and the chance that he will dial all six correctly is $1/10$ multiplied itself 6 times. This is $(1/10)^6 = 1/10^6 = 1/1,000,000$. Choice F is incorrect because $1/7$ assumes there are only 7 possible answers. Choice G is incorrect because $1/10$ is the chance that Manolo will dial any one of the digits correctly. Choice I is incorrect because $1/10,000,000$ is the chance of dialing a standard seven-digit number purely by guessing, but Manolo only has to guess six of these digits.

37 Which of the following CANNOT be assumed about the parallelogram shown below?

Analysis: *Choice D is correct.* It cannot be assumed that $BC \perp CD$ because ABCD is a parallelogram, not a square or a rectangle. A parallelogram has two sets of equal sides but the angles are not necessarily 90°. Choice A is incorrect because the sum of the interior angles of a parallelogram is 360°.

Choice B is incorrect because opposite sides of a parallelogram are parallel. Choice C is incorrect because in all parallelograms opposite sides are congruent. Thus for any parallelogram the sum of two adjacent sides must be equal to the sum of the other two adjacent sides.

38 A clothing company is designing a symbol to use as its logo. The logo is made up of a square, two equilateral triangles, and two semicircles. In the diagram of the logo shown below, you are given the length of the diagonal of the square and the height of one of the triangles. What is the area of the logo? Round your answer to the nearest **tenth**.

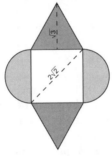

Analysis: *The answer is approximately 10.6.* The key to finding the area of each of the three shapes that form this logo is the length of the square's side. When we know that, we will be able to find the area of the square ($A = lw$), because $l = w$ in a square. We will also know the base of each equilateral triangle, and since we already know its height, we will be able to find its area ($A = \frac{1}{2} bh$). We will also know the diameter of the circle (the two semi-circles make 1 whole circle), and will know that its radius is half its diameter so we can find the circle's area ($A = \pi r^2$). Let each square's side

equal x. Apply the Pythagorean Theorem remembering that the lone variable, c^2, is the hypotenuse (longest side) of the right triangle. $a^2 + b^2 = c^2$; $x^2 + x^2 = (2\sqrt{2})^2$; $2x^2 = (2^2)(2\sqrt{2})^2$; $2x^2 = 4(2)$; $2x^2 = 8$; $x^2 = 4$; $x = 2$. So the square's sides are 2 and the area of the square is $A = 2 \times 2 = 4$. The area of each equilateral triangle is $A = \frac{1}{2} bh = \frac{1}{2}(2)(\sqrt{3}) = \frac{1}{2} 2\sqrt{3} = \sqrt{3}$. So the area of both triangles is $2\sqrt{3}$. The radius of the circle is 1, so the area of the circle is $A = \pi r^2 = \pi 1^2 = \pi$. The area of the logo is the area of the square plus the area of the triangles plus the area of the circle or: $4 + 2\sqrt{3} + \pi \approx 4 + 3.46 + 3.14 \approx 10.6$

39 Hortense is increasing the size of her garden. The original dimensions of the garden are 20 yards by 25 yards. If the garden is lengthened by 10 yards on each side, how will this affect the perimeter of the garden?
Analysis: *Choice G is correct.* If the length of each side of the garden is increased by 10 yards, the perimeter of the garden will increase by 10 yards x 4 = 40 yards. Choice F is incorrect because the perimeter of the garden would increase by 80 yards if the length of each side were increased by 20 yards. Choice H is incorrect because perimeter of the garden would increase by 20 yards if the length of each side were increased by 5 yards. Choice I is incorrect because the length of all four sides of the garden are increased by 10 yards, not just one of the sides.

Mathematics Assessment One: Answer Key

40 Which of the following represents the **greatest** number?
Analysis: *Choice D is correct.* The number $(3)^{-2}$ is equivalent to $1/9$, or approximately 0.11. Choice A is incorrect because $(0.25)^3$ is equivalent to $0.25 \times 0.25 \times 0.25 = 0.015625$. This is less than 0.11. Choice C is incorrect because $(2)^{-5}$ is equivalent to $1/(2 \times 2 \times 2 \times 2 \times 2) = 1/32$, or 0.03125. This is less than 0.11. Choice B is incorrect because $(-1)^5$ is equivalent to $-1 \times -1 \times -1 \times -1 \times -1 = -1$. This is less than 0.11.

41 If triangle ABC were translated such that point B's new coordinates were (6, 0), what would be the new coordinates for point C?

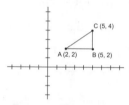

Analysis: *Choice H is correct.* Points B and C are on the same vertical line, they will always have the same *x*-coordinate. If the new *x*-coordinate of point B is a 6, then point C's new *x*-coordinate must be a 6. The original *y*-coordinates are two units apart, the new *y*-coordinates must also be 2 units apart. You can also solve this by noticing that when point B is translated, it moves right one unit and down 2 units. This same translation also applies to every other point in the figure, so C moves right one unit $(5 + 1 = 6)$ and down 2 units $(4 - 2 = 2)$.

42 Chip and Jen are designing a birdhouse for a school project. For the roof and base of the birdhouse, they used pieces of wood measuring 6 inches by 6 inches by 0.5 inch. For the four sides of the birdhouse, they used pieces of wood measuring 8 inches by 6 inches by 0.5 inch. After they had assembled these pieces into a rectangular prism, they added a small piece of wood measuring 1

inch by 3 inches by 0.5 inch to serve as a ledge. To complete the project, they cut a cylinder with a diameter of 2 inches out of the front of the birdhouse for the birds to enter. What is the total volume of wood in the completed birdhouse? Your answer should be in cubic inches rounded to the nearest tenth.

Analysis: *The correct answer is 131.9 in^3.* Find the volume of the roof $(6 \times 6 \times 0.5 = 18\ in^3)$. Since the base is the same size, its volume is also 18 cubic inches. Find the volume of each of the four sides $(6 \times 8 \times 0.5 = 24\ in^3)$. Since there are four, the total volume of the sides is 96 cubic inches. Now you must add the value for the ledge they added to the birdhouse. The volume of the ledge is $1 \times 3 \times 0.5 = 1.5\ in^3$. Subtract the volume of the hole using the formula for the volume of a cylinder $(V = \pi r^2 h; V = \pi 1^2(0.5) = 1.57$ cubic inches). Combine all of the information you have found: $18 + 18 + 96 + 1.5 - 1.57 = 131.93$ cubic inches. Rounded to the nearest tenth, the correct answer is 131.9 cubic inches.

43 A common garden snail travels at a speed of 0.03 miles per hour. The giant tortoise travels at a speed of 0.17 miles per hour. Both the snail and the tortoise took the same amount of time to travel from their respective starting points to the tree. If the tortoise and the snail both arrive at a tree at the exact same moment, how far away from the tree was the garden snail when the tortoise was 2 miles away from the tree? Round to the nearest **hundredth.**

Analysis: *The correct answer is 0.35 miles away.* Use the information about the distance the tortoise traveled as the key to this problem: 2 miles at a speed of 0.17 mph. The distance formula is Distance equals Rate times Time or $d = rt$. Substituting in the tortoise's numbers we get: $2 = 0.17 \times t$. Solve for *t*, we get $t = 2/0.17$ so $t = 11.76$ hours. So it took the tortoise 11.76 hours to get to the tree. Plug the time into the rate formula for the snail, $d = rt$; $d = 0.03 \times 11.76 = 0.35$ miles. The snail was 0.35 miles from the tree.

44 If pentagon ABCDE is reflected across the *y*-axis, what are the new coordinates of point D?

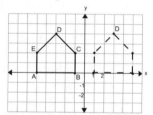

Analysis: *Choice A is correct.* When reflecting an object across a line, each point in the reflected image must be the same distance from the line of reflection as the original. In this case the line of reflection is the *y*-axis, so each point in the reflected pentagon must be the same distance from the *y*-axis as the original pentagon. Notice that in the original pentagon, Point D is at (-3, 4). When it's reflected across the *y*-axis, its *y*-coordinate will not change. It will still be 4. However, to make the reflection of Point D 3 units from the *y*-axis, its *x*-coordinate will have to change from negative to positive. The *x*-coordinate must be 3. It may help you to sketch the pentagon after it is reflected and then determine the coordinates of the new points.

Mathematics Assessment One: Answer Key

45 Using the Pythagorean theorem, what is the distance between C and A?

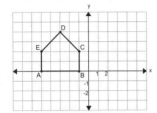

Analysis: *Choice G is correct.* Notice that Points A, B, and C form a right triangle, so we can apply the Pythagorean Theorem. We can determine the length of both legs of this triangle from the diagram. We need to find the hypotenuse, the triangle's longest side, and the lone variable, c, in the formula. Line segment CB's length is 2 and line segment BA's length is 4. Plug these values into the Pythagorean theorem: $a^2 + b^2 = c^2$; $2^2 + 4^2 = c^2$; $4 + 16 = c^2$; $c^2 = 20$; $C = \sqrt{20}$. Notice that $\sqrt{20}$ is not one of your answer choices, so try to find factors of 20 that are perfect squares: $C = \sqrt{20} = \sqrt{4 \times 5} = \sqrt{4}\ \sqrt{5} = 2\sqrt{5}$

46 Which of the following statements best describes the combined area of triangles ABC and AZY?

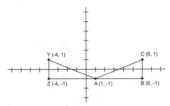

Analysis: Choice B is correct. The area of one triangle is $A = 1/2bh = 1/2(5)(2) = 5$, but the other triangle is congruent to the first so it's area is also 5. Their combined area is $5 + 5 = 10$. This is the same area as a rectangle with a length (base) = 5 and a width (height) = 2 (2 x 5 = 10). If you reflect one of the triangles in your mind and translate it on top of the other you can see that together they form a 2 by 5 rectangle. Choice A is incorrect since the hypotenuse is not needed for solving for area and the quadrant makes no difference to the area calculation. Choice C is incorrect since the lengths can be determined from the coordinates provided. Choice D is incorrect; where the base is located is irrelevant because you can determine the lengths of the bases from the illustration.

47 Which of the following pictures best demonstrates the basic premise of the Pythagorean theorem, using area to demonstrate the reason the theorem works?
Analysis: *Choice H is correct.*

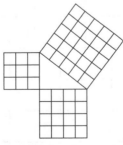

This answer choice demonstrates why the sum of the legs squared equals the square of the hypotenuse: $3^2 + 4^2 = 5^2$, $9 + 16 = 25$.

48 Dean wants to know how tall a tree in his yard is, but he does not have the proper tools to measure it by hand. He does know that he is 6 feet tall, and he is able to measure the length of his shadow (12 feet) and the length of the shadow of the tree (36 feet). How tall is the tree in inches?
Analysis: *The correct answer is 216 inches.* To find the actual value of the height of the tree, set up a proportion using the known lengths: 6 feet/12 feet = tree height/36 feet; cross-multiplying yields a tree height of 18 feet. Don't forget to convert to inches. The tree height is 216 inches. (18 x 12 = 216)

49 Drive-Away-Today auto dealership sells a car for $1,400.00 dollars above its wholesale cost. The manufacturer will also give the dealership a $200.00 incentive fee for each car that is sold. Which of the following functions describes the total profit, $f(c)$, for number of cars sold, c?
Analysis: *Choice C is correct.* The function that describes this situation is $f(c) = \$1,400c + \$200c$. Simplified, this is $f(c) = \$1,600c$.

50 Brenda received $120,000.00 when a relative's estate was settled. She wants to invest this money. She plans to follow the advice of a financial advisor who has told her to invest 50% of this money in a savings account, 20% in bonds, and the remaining amount in stocks. How much money will be invested in stocks?
Analysis: *Choice H is correct.* If Brenda follows the advice she is given, 30% of the money will go to stocks ($120,000 − 50% (savings) − 20% (bonds) = 30% (stocks)). (0.30 x $120,000.00 = $36,000.00)

Mathematics Assessment One: Correlation Chart

The Correlation Charts can be used by the teachers to identify areas of improvement. When students miss a question, place an "X" in the corresponding box. A column with a large number of "Xs" shows more practice is needed with that particular standard.

Correlation	MA.E.3.4.1	MA.A.1.4.3	MA.D.1.4.1	MA.E.2.4.2	MA.B.3.4.1	MA.A.4.4.1	MA.D.1.4.1	MA.E.1.4.2	MA.D.1.4.1	MA.C.1.4.1	MA.E.2.4.1	MA.D.1.4.2	MA.D.2.4.1	MA.C.2.4.1	MA.A.3.4.1	MA.C.3.4.1	MA.E.1.4.2	MA.A.2.4.2	MA.B.2.4.2	MA.A.1.4.1
Answer	A	I	D	G	C	I	*	C	G	B	*	*	F	C	H	*	C	F	A	G
Question	1	2	3	4	5	6	7	8	9	10	11	12	13	14	15	16	17	18	19	20

Student Names

*Gridded-Response Item

Mathematics Assessment One: Correlation Chart

Correlation	MA.D.1.4.1	MA.B.2.4.1	MA.E.1.4.1	MA.D.1.4.1	MA.C.3.4.2	MA.B.1.4.1	MA.D.1.4.1	MA.C.3.4.2	MA.E.1.4.3	MA.A.3.4.3	MA.C.2.4.1	MA.D.1.4.1	MA.D.2.4.1	MA.D.1.4.1	MA.A.3.4.1	MA.E.2.4.2	MA.C.1.4.1	MA.B.1.4.1	MA.D.1.4.2	MA.A.1.4.2
Answer	C	*	*	H	C	F	D	H	*	A	H	B	G	*	B	H	D	*	G	D
Question	21	22	23	24	25	26	27	28	29	30	31	32	33	34	35	36	37	38	39	40

Student Names

*Gridded-Response Item

Mathematics Assessment One: Correlation Chart

Correlation	MA.C.2.4.1	MA.B.1.4.1	MA.B.2.4.2	MA.C.2.4.1	MA.C.3.4.1	MA.C.3.4.2	MA.C.1.4.1	MA.B.1.4.3	MA.D.1.4.1	MA.A.1.4.4
Answer	H	*	*	A	G	B	H	*	C	H
Question	41	42	43	44	45	46	47	48	49	50

Student Names

* Gridded-Response Item

Mathematics Assessment Two

Directions for Taking the Mathematics Assessment Two

On this section of the Florida Comprehensive Assessment Test (FCAT), you will answer 50 questions. For multiple-choice questions, you will be asked to pick the best answer out of four possible choices and fill in the answer in the answer bubble. On gridded-response questions, you will also fill your answer in answer bubbles, but you will fill in numbers and symbols corresponding to the solution you obtain for a question. Fill in the answer bubbles and gridded-response answer bubbles on the Answer Sheet on pages 205–209 to mark your selection.

Read each question carefully and answer it to the best of your ability. If you do not know an answer, you may skip the question and come back to it later.

Figures and diagrams with given lengths and/or dimensions are not drawn to scale. Angle measures should be assumed to be accurate. Use the formula sheet on page 33 and the conversion chart on page 34 to help you answer the questions. You will also be given a calculator to use.

When you finish, check your answers.

1 The following table lists the masses of four of the moons of Jupiter.

Moon of Jupiter	Mass (in kg)
Callisto	1.08×10^{23}
Europa	4.80×10^{22}
Ganymede	1.48×10^{23}
Io	8.93×10^{22}

Which moon has the **largest** mass?

A. Callisto
B. Europa
C. Ganymede
D. Io

2 Which of the choices below shows the numbers ordered from **largest** to **smallest**?

F. $3, \sqrt{3}, \frac{1}{\sqrt{3}}, \frac{1}{3}$

G. $\sqrt{3}, 3, \frac{1}{3}, \frac{1}{\sqrt{3}}$

H. $3, \sqrt{3}, \frac{1}{3}, \frac{1}{\sqrt{3}}$

I. $\sqrt{3}, 3, \frac{1}{\sqrt{3}}, \frac{1}{3}$

3 Which of the choices below shows the same number represented in three equivalent forms?

    A. $\frac{1}{25}$, 0.25, 25%

    B. $\frac{1}{4}$ 0.25, 2.5%

    C. $\frac{4}{10}$, 0.04, 40%

    D. $\frac{1}{25}$, 0.04; 4%

4 Which of the following numbers is NOT equivalent to $\frac{1}{(-5)^2}$?

    F. $\frac{1}{5^2}$

    G. $\frac{1}{5^{-2}}$

    H. 5^{-2}

    I. $(-5)^{-2}$

Go On

5 Cheryl receives a monthly allowance of $25.00. She also has a part-time job that pays $6.50 per hour, but she must pay $32.50 each month to ride the bus between her home and her job. After subtracting off her transportation costs, Cheryl deposits half of her earnings into her savings account every month. If she works w hours during the month, the amount of money she deposits into her savings account, S, can be represented by the formula below:

$$S = \tfrac{1}{2}(25 + 6.5w - 32.5)$$

How much money will Cheryl add to her savings in September if she works 36 hours?

6 Mr. Ruiz bought three books during his local bookstore's summer sale. Two of the books originally cost $12.95 and the third book originally cost $9.95. All three books were marked 10% off. He also paid a 6% sales tax. How much money did Mr. Ruiz spend altogether? Round your answer to the nearest penny.

7 The bar graph below shows the population of each of four metropolitan areas in Florida.

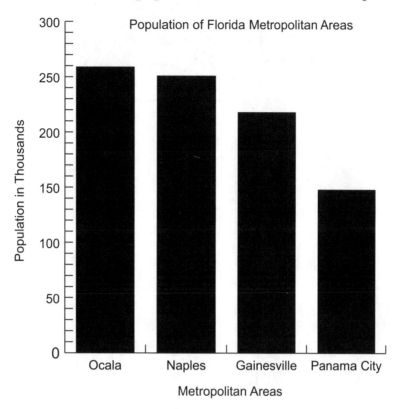

Which of the choices below is a good estimate of the total population of these four metropolitan areas?

A. 900,000
B. 1,000,000
C. 1,100,000
D. 1,200,000

8 Becky's school has arched doorways with the dimensions shown below.

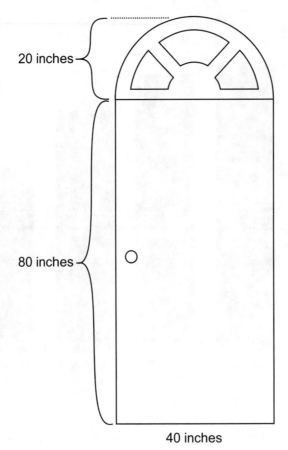

Which of the expressions below can be used to calculate the area of the doorway?

F. $2(80 + 40) + \pi 20^2 \times \frac{1}{2}$

G. $80 \times 40 + \pi 20^2 \times \frac{1}{2}$

H. $2(80 + 40) + \pi 40^2 \times \frac{1}{2}$

I. $80 \times 40 + \pi 20^2$

9 What is the measure in inches of the curved edge of one section of the dartboard shown below? Round your answer to the nearest **hundredth**.

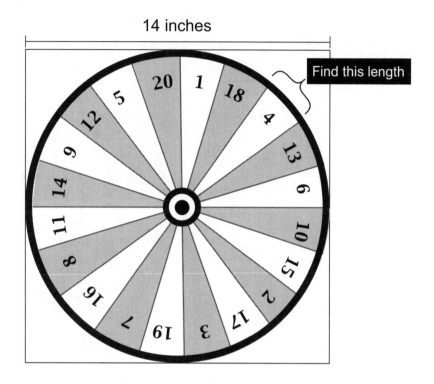

10 A map shows the distance between Jacksonville and Orlando as 7.0 centimeters and the distance between Orlando and Fort Lauderdale as 10.6 centimeters.

If the actual distance between Jacksonville and Orlando is 140.4 miles, what is the actual distance between Orlando and Fort Lauderdale? Round your answer to the nearest **tenth** of a **mile**.

© **Englefield & Associates, Inc.**

11 Linda wants to fill this rectangular sandbox two-thirds full of sand.

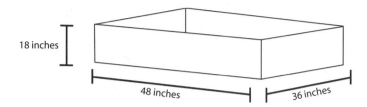

18 inches

48 inches

36 inches

Linda can buy sand in bags that hold 1.5 cubic feet each. How many bags of sand will she need?

A. 6

B. 8

C. 12

D. 18

12 Rhombus ABCD is shown in the graph below.

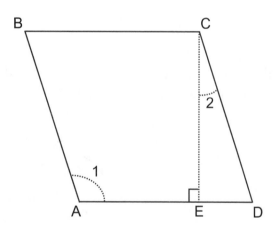

If ∠1 measures 127°, what is the degree measure of ∠2?

13 Parallelogram LMNO is shown below.

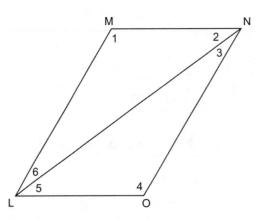

Which of the following equations is NOT true for the measures of angles 1, 2, 3, 4, 5, and 6?

F. $m\angle 1 + m\angle 2 = m\angle 4 + m\angle 5$
G. $m\angle 1 + m\angle 6 = m\angle 4 + m\angle 3$
H. $m\angle 4 + m\angle 5 = m\angle 1 + m\angle 6$
I. $m\angle 3 + m\angle 5 = m\angle 2 + m\angle 6$

14 Triangle QRS is inscribed in a circle centered at point P as shown in the sketch below. The radius of circle P is 3 centimeters and segment SP is perpendicular to segment QR.

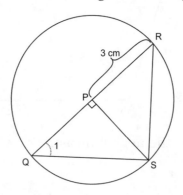

What is the degree measure of $\angle 1$?

Go On

15 Triangle QRS is inscribed in a circle centered at point P as shown in the sketch below. The radius of circle P is 3 centimeters and segment SP is perpendicular to segment QR.

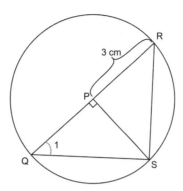

What is the length in centimeters of segment QS? Express your answer as a decimal rounded to the nearest **hundredth**.

16 Triangle ABC is shown below. Suppose there is a triangle DEF that is similar to triangle ABC. The shortest side of triangle DEF is segment DF, which measures 10 cm.

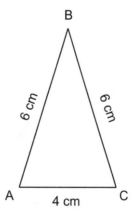

What is the perimeter of triangle DEF in **centimeters**?

17 ABC and DEF are similar right triangles and segment AC corresponds to segment DF as shown below.

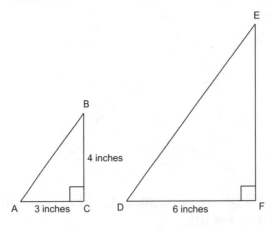

Which of the following statements is NOT true?

A. The length of side DE is 10 inches.
B. The area of triangle DEF is 12 square inches.
C. The perimeter of triangle DEF is 24 inches.
D. The measure of angle BAC equals the measure of angle EDF.

18 Quadrilateral JKLM is shown in the graph below.

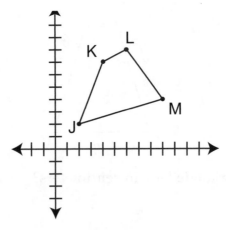

Suppose JKLM is reflected once over the x-axis and then translated by –2 along the y-axis. What are the new coordinates point M?

F. (9, –6)
G. (–11, 4)
H. (9, –2)
I. (–9, 2)

19 JKL is a triangle with the coordinates shown below.

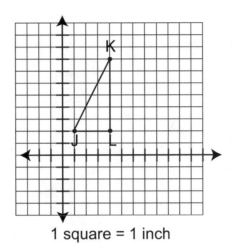

1 square = 1 inch

What is the length of segment JK in **inches**? Round your answer to the nearest **hundredth**.

20 The graph of line ℓ is shown below.

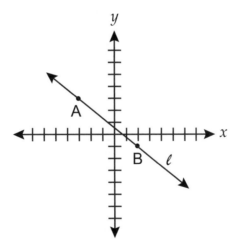

What is the slope of line ℓ?

A. −0.8
B. −1.25
C. 0.8
D. 1.25

21 The swimming pool at Northland Swim Club is a parallelogram and is drawn to scale in the graph below.

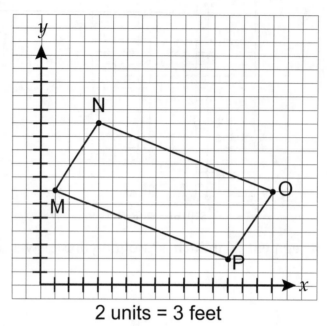

2 units = 3 feet

What is the perimeter of the actual swimming pool in **feet**?

F. 30 ft
G. 32 ft
H. 48 ft
I. 54 ft

Go On ▶

22 The function table below shows values of x and y that satisfy the equation $-2x^3 + y = 0$.

x	y
0	0
1	2
2	16
3	54
4	128
5	

What is the value of y when $x = 5$?

23 For the function table shown below, what is the value of y when $x = 5$?

x	y
0	1
1	4
2	13
3	28
4	49
5	
6	109

A. 70
B. 73
C. 76
D. 77

Go On

24 The Turners pay a monthly fee of $34.99 for their phone service. For this monthly fee, they can make 300 minutes' worth of phone calls at no additional charge. For every minute over 300, they pay an additional $0.40. Assuming that the Turners use their phone service for more than 300 minutes each month, which of the following equations shows P, the Turners' total phone bill, as a function of m, the total number of minutes?

F. $P = 0.40m - 85.01$
G. $P = 0.40m + 85.01$
H. $P = 0.40m + 154.99$
I. $P = 0.40m - 154.99$

25 Rikkhyia wants to calculate the cost of driving her car to work for one week. She works five days each week and the distance between her home and her workplace is 14.1 miles. Her car averages 30 miles per gallon of gasoline. In addition to paying for gas, she must also pay $4.00 per day for parking. Let g equal the cost of a gallon of gasoline. Which of the equations below shows Rikkhyia's total cost, C, as a function of g?

A. $C = \frac{30}{141} \times g + 20$

B. $C = \frac{141}{30} \times g + 20$

C. $C = \frac{141}{30} \times g + 4$

D. $C = \frac{30}{141} \times g + 4$

26 Mary Ann is preparing to move, and she is packing her clothes into boxes that measure 12 inches by $15\frac{1}{2}$ inches by 10 inches. She wants to find new boxes that will allow her pack four times the volume of clothing in each box. Of the boxes described below, which one should Mary Ann NOT choose?

F. 12 inches x $15\frac{1}{2}$ inches x 40 inches

G. 24 inches x 31 inches x 20 inches

H. 24 inches x $15\frac{1}{2}$ inches x 20 inches

I. 12 inches x 31 inches x 20 inches

27 Janice makes lemonade using lemons, sugar, and water. She uses the juice from 4 lemons and 3 cups of sugar for every gallon of lemonade. If she uses 18 lemons altogether, how many cups of sugar should she use? Express your answer as a **fraction or** a **decimal**.

28 Karen runs for 45 minutes every morning. She averages 5.5 miles a day. If she wants to run an average of 8 miles each day, by how much does she need to increase her average speed?

A. $2\frac{1}{2}$ miles/hour

B. $7\frac{1}{3}$ miles/hour

C. $10\frac{2}{3}$ miles/hour

D. $3\frac{1}{3}$ miles/hour

29 The equation $y = 2x + 2$ is graphed below.

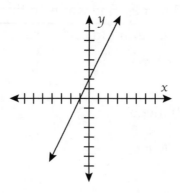

Which of the following graphs shows this equation with the first 2 replaced by a 3?

F.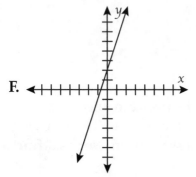

H.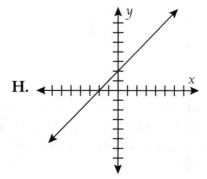

G.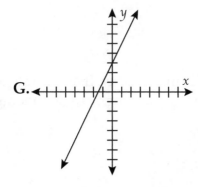

I.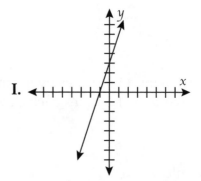

Go On

30 Alex and John are both walking to school. They have agreed to meet on the school grounds as soon as they arrive. Alex's home is 0.6 mile from the school and he walks an average of 4 miles per hour. John's home is 0.9 mile away and he walks an average of 4.5 miles per hour. If they leave home at the same time, how many minutes will one of the boys wait for the other?

31 Isaac is going to the state fair. He can either pay $12.00 for admission plus $1.75 for every ride, or he can pay $30.00 for admission and ride for free all day. What is the maximum number of rides Isaac can go on and still pay less with the $12.00 admission price?

A. 6
B. 10
C. 11
D. 17

32 Jonathan is painting 6 circular columns on the front porch of an apartment building. For each column, he can use the formula $S.A. = 2\pi rh$ to calculate the area of a column's exposed surface that he will need to paint. Neither the upper nor the lower base of the column are visible, so they will not be painted. The height of each column is 9.5 feet and the diameter is 3 feet. The paint he is using will cover 350 square feet per gallon, but it is sold by the quart. How many **quarts** of paint should he buy?

33 The Argosy Bookshop has the following inventory:

Types of Books	Number of Books
Fiction & Poetry	9,042
History & Biography	10,118
Travel	5,431
Other Nonfiction	4,087

At the end of a week, they have sold 270 books in the fiction and poetry category. If they've sold the same percentage of travel books, how many travel books have they sold? Round your answer to the nearest **whole number**.

34 Kevin was interested in the different combinations of pets owned by his classmates. He conducted a survey in which he asked each student whether or not they owned a dog, a cat, or a fish. He represented his findings with the Venn diagram shown below, in which Circle A is the set of all dog owners, Circle B is the set of all cat owners, and Circle C is the set of all fish owners.

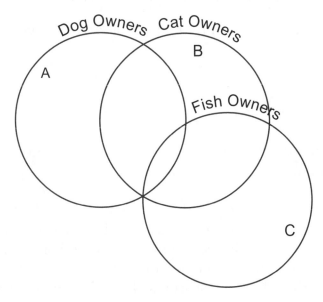

Which of the following statements is true for the students in Kevin's class?

F. No one owns a dog, a cat, and a fish.
G. No one owns a dog and a cat but not a fish.
H. No one owns a dog and a fish but not a cat.
I. No one owns a cat and a fish but not a dog.

35 The heights of 14 people are shown in the line plot below.

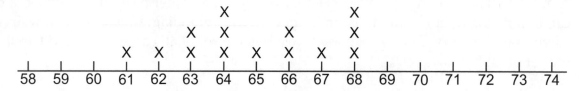

What is the median height for this group of people? Express your answer in **inches**.

36 Suppose you were told that the mean of the set of all test scores for a class of 28 students is 76. Which of the following statements is true?

A. Multiplying 76 by 28 will give you the total of all the students' test scores.
B. Of the 28 students in the class, 14 scored at or above 76 and 14 scored at or below 76.
C. If a student in this class is selected at random, the probability that they had a test score of 76 is as high or higher than the probability of any other score.
D. If r is the range of the set of all test scores, then all of the scores are between the numbers $76 - r/2$ and $76 + r/2$.

37 Jay has signed up to play baseball at his neighborhood recreation center. He will be assigned to one of the following teams: Marlins, Astros, Diamondbacks, or Dodgers. The coach will assign him to play one of the following positions on the team: first baseman, second baseman, third baseman, pitcher, catcher, shortstop, right fielder, center fielder, or left fielder.

Teams
Marlins
Astros
Diamondbacks
Dodgers

Positions
first baseman
second baseman
third baseman
pitcher
catcher
shortstop
right fielder
center fielder
left fielder

If the assignments are made at random, what is the probability that Jay will be a fielder for the Marlins? Express your answer as a **fraction**.

38 Suppose you are playing a board game that uses two spinners and a die to determine a player's next move. A player spins both arrows and rolls the die, and then moves ahead by the number of spaces shown on the die only if both arrows land on black. What are the chances that you will move ahead on your next turn?

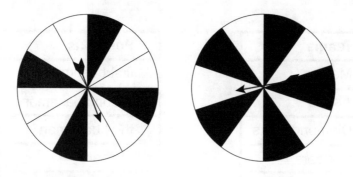

  F. 0.139
  G. 0.167
  H. 0.409
  I. 0.833

39 Every student at Fort Hayes High School takes one, and only one, foreign language class. The following table shows the number of students who study each of the languages offered at Fort Hayes.

	Spanish	**French**	**Latin**	**Japanese**
9th Grade	127	62	27	13
10th Grade	102	71	33	25
11th Grade	82	79	45	38
12th Grade	90	57	75	40

What is the probability that an 11th grader chosen at random is taking French? Express your answer as a **percent** rounded to the nearest whole number.

40 Consolidated Fish Company has a pier-side location where it buys shrimp directly from the fishing fleet. It pays $2.25 per pound for shrimp. Usually the shrimp are brought to market in 40-pound capacity plastic containers. Which of the following represents a formula that could be used for buying any quantity of shrimp, where P is the overall amount paid for a catch and S is the weight of the shrimp being sold?

    A. $P = \$2.25(40)$
    B. $P = \$2.25(S)$
    C. $P = \$2.25(S) + 40$
    D. $P = S(40)$

41 The process of simplifying the equation $2y = 2x + 2$ best demonstrates which of the following properties?

    F. commutative
    G. associative
    H. distributive
    I. inverse

42 In mathematics, what is a dilation?

    A. the "flipping" of a figure over an imaginary line of symmetry
    B. the reducing or enlarging of a figure in proper proportion
    C. the "sliding" of a figure to a new location on the same coordinate plane
    D. the rotation of a figure around some fixed point

43 Jane is preparing to mail an important set of design plans. She has rolled the plans up and placed them inside a cylinder-shaped tube that is 27 inches tall and has a radius of 2 inches. Jane plans to select a rectangular box that will allow at least 2 inches of packing material on all sides of the plans. What is the smallest box she can use for mailing the plans?

F. 6 inches by 6 inches by 29 inches
G. 8 inches by 8 inches by 29 inches
H. 8 inches by 8 inches by 31 inches
I. 10 inches by 10 inches by 31 inches

44 Mason dug a trench 60 feet long, 2 feet wide, and 2 feet deep as part of a construction project. This soil will have to be removed in pickup truck loads until it has been completely cleared from the property. The pickup truck bed will safely carry a load of soil that measures 4 feet by 8 feet by 2 feet. How many truckloads will have to be made to clear the soil from the property?

A. 3
B. 4
C. 5
D 8

45 A large rectangular-shaped field measures 600 feet by 800 feet. How long would a fence need to be if it were installed from one corner of this field to the opposite diagonal corner?

F. 700
G. 800
H. 1,000
I. 1,200

46 A ladder that is eight feet tall has been leaned against a wall at an angle, and the base of the ladder is firmly secured at a point exactly three feet from the wall. Rounded to the nearest foot, how high up the wall is the top of the ladder?

A. 7
B. 8
C. 9
D. 10

47 Rene sees a box on a shelf and notices its height is about 12 inches and its width is about 10 inches. On the end of the box that she can see, it is labeled 3 cubic feet of storage, but the rest of the box extends back into the shelf and cannot be seen. About how long should the box be if 3 cubic feet is its actual volume?

F. 12 inches
G. 20 inches
H. 43 inches
I. 100 inches

48 The floor plan of a new office complex calls for 35,000 total square feet. Construction costs for this facility are estimated to be $122.00 per square foot, based on the design plans. How much money will this office complex cost to build if the estimate is correct?

A. $4,000,000
B. $4,270,000
C. $4,720,000
D. $4,900,000

Go On

49 What is the probability that a black marble in a bag with 3 blue marbles will be drawn on the first attempt, and that a coin flipped simultaneously with the marble drawing will land on heads instead of tails?

    F. 0.10
    G. 0.125
    H. 0.25
    I. 0.555

50 On the last test in Mr. Counter's math class, the median score was an 87 and the mode was 92. Harold argued that this meant the mean score must be higher than 87. Is Harold correct?

    A. Harold is correct because the mean must be greater than the middle number if the number occurring most often is greater than the middle number.
    B. Harold is correct because the mean must be less than the middle number if the number occurring most often is less than the middle number.
    C. Harold is incorrect because the mean must be less than the middle number if the number occurring most often is greater than the middle number.
    D. Harold is incorrect because there is not enough information to determine the mean.

This is the end of Mathematics Assessment Two.
Until time is called, go back and check your work or answer
questions you did not complete. When you have finished, close
your workbook.

BLANK PAGE

Name _____

Answer all the questions that appear in the Mathematics Assessment Two on this Answer Sheet.

1 Ⓐ Ⓑ Ⓒ Ⓓ   **2** Ⓕ Ⓖ Ⓗ Ⓘ   **3** Ⓐ Ⓑ Ⓒ Ⓓ

4 Ⓕ Ⓖ Ⓗ Ⓘ   **5**   **6**

7 Ⓐ Ⓑ Ⓒ Ⓓ   **8** Ⓕ Ⓖ Ⓗ Ⓘ   **9**

Fold and Tear Carefully Along Dotted Line.

10
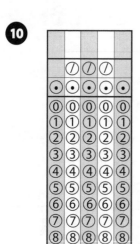

11　Ⓐ　Ⓑ　Ⓒ　Ⓓ

12
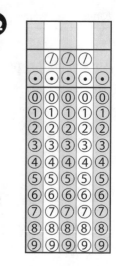

13　Ⓕ　Ⓖ　Ⓗ　Ⓘ

14

15

16

17　Ⓐ　Ⓑ　Ⓒ　Ⓓ

18　Ⓕ　Ⓖ　Ⓗ　Ⓘ

Fold and Tear Carefully Along Dotted Line.

19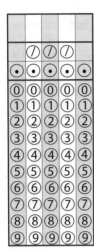

20 Ⓐ Ⓑ Ⓒ Ⓓ   **21** Ⓕ Ⓖ Ⓗ Ⓘ

22

23 Ⓐ Ⓑ Ⓒ Ⓓ   **24** Ⓕ Ⓖ Ⓗ Ⓘ

25 Ⓐ Ⓑ Ⓒ Ⓓ   **26** Ⓕ Ⓖ Ⓗ Ⓘ   **27**

28 Ⓐ Ⓑ Ⓒ Ⓓ　　　　**29** Ⓕ Ⓖ Ⓗ Ⓘ

30

31 Ⓐ Ⓑ Ⓒ Ⓓ

32 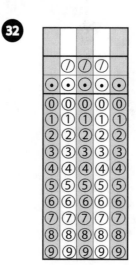　　　**33**

34 Ⓕ Ⓖ Ⓗ Ⓘ

35 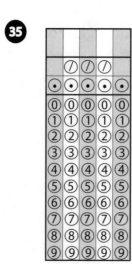　　　**36** Ⓐ Ⓑ Ⓒ Ⓓ

Fold and Tear Carefully Along Dotted Line.

37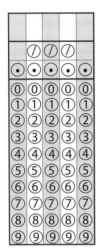

38 Ⓕ   Ⓖ   Ⓗ   Ⓘ

39

40 Ⓐ   Ⓑ   Ⓒ   Ⓓ

41 Ⓕ   Ⓖ   Ⓗ   Ⓘ

42 Ⓐ   Ⓑ   Ⓒ   Ⓓ

43 Ⓕ   Ⓖ   Ⓗ   Ⓘ

44 Ⓐ   Ⓑ   Ⓒ   Ⓓ

45 Ⓕ   Ⓖ   Ⓗ   Ⓘ

46 Ⓐ   Ⓑ   Ⓒ   Ⓓ

47 Ⓕ   Ⓖ   Ⓗ   Ⓘ

48 Ⓐ   Ⓑ   Ⓒ   Ⓓ

49 Ⓕ   Ⓖ   Ⓗ   Ⓘ

50 Ⓐ   Ⓑ   Ⓒ   Ⓓ

Fold and Tear Carefully Along Dotted Line.

BLANK PAGE

Mathematics Assessment Two: Skills Chart

Question	Standard	Answer	Keywords
1	MA.A.1.4.2	C	Understands different ways numbers are represented
2	MA.A.1.4.2	F	Understands different ways numbers are represented
3	MA.A.1.4.4	D	Understands that numbers can be represented in a variety of equivalent forms
4	MA.A.3.4.1	G	Understands and explains the effects of operations on numbers
5	MA.A.3.4.3	$113.25	Selects and justifies alternate strategies
6	MA.A.3.4.3	$34.20	Uses appropriate methods of computing
7	MA.A.4.4.1	A	Uses estimation to predict
8	MA.B.1.4.1	G	Uses graphic models to derive formulas for finding area
9	MA.B.1.4.2	2.20	Uses graphic models to derive formulas for finding angle measures and arc lengths
10	MA.B.1.4.3	212.6	Relates concepts of measurement to real-world problems
11	MA.B.2.4.1	B	Selects and uses appropriate measurements
12	MA.C.1.4.1	37°	Uses properties of geometric shapes to const formal and informal proofs
13	MA.C.1.4.1	H	Uses properties of geometric shapes to const formal and informal proofs
14	MA.C.1.4.1	45°	Uses properties of geometric shapes to const formal and informal proofs
15	MA.C.1.4.1	4.24	Uses properties of geometric shapes to const formal and informal proofs
16	MA.C.2.4.1	40	Understands geometric concepts
17	MA.C.2.4.1	B	Understands geometric concepts
18	MA.C.3.4.1	F	Represents and applies geometric properties to solve real-world problems
19	MA.C.3.4.1	6.71	Represents and applies geometric properties to solve real-world problems
20	MA.C.3.4.2	A	Uses a graph and applies properties of two- and three-dimensional figures
21	MA.C.3.4.2	I	Uses a graph and applies properties of two- and three-dimensional figures
22	MA.D.1.4.1	250	Describes, analyzes, and generalizes functions
23	MA.D.1.4.1	C	Describes, analyzes, and generalizes functions
24	MA.D.1.4.1	F	Describes, analyzes, and generalizes functions
25	MA.D.1.4.1	B	Describes, analyzes, and generalizes functions

Mathematics Assessment Two: Skills Chart

Question	Standard	Answer	Keywords
26	MA.D.1.4.2	G	Determines impact when changing parameters of given functions
27	MA.D.1.4.2	13.5	Determines impact when changing parameters of given functions
28	MA.D.1.4.2	D	Determines impact when changing parameters of given functions
29	MA.D.1.4.2	F	Determines impact when changing parameters of given functions
30	MA.D.2.4.2	3	Uses systems of equations and inequalities to solve real-world problems
31	MA.D.2.4.2	B	Uses systems of equations and inequalities to solve real-world problems
32	MA.D.2.4.2	7	Uses systems of equations and inequalities to solve real-world problems
33	MA.D.2.4.2	162	Uses systems of equations and inequalities to solve real-world problems
34	MA.E.1.4.1	H	Interprets data
35	MA.E.1.4.2	64.5	Calculates measures of central tendency (mean, median, and mode)
36	MA.E.1.4.2	A	Calculates measures of central tendency (mean, median, and mode)
37	MA.E.2.4.1	1/12	Determines probabilities
38	MA.E.2.4.1	G	Determines probabilities
39	MA.E.3.4.1	32	Designs and performs experiments involving more than one variable
40	MA.D.1.4.1	B	Describes, analyzes, and generalizes functions
41	MA.A.3.4.2	H	Selects and justifies alternative strategies
42	MA.C.2.4.1	B	Understands geometric concepts
43	MA.B.1.4.1	H	Measures quantities in the real world and uses the measures to solve problems
44	MA.B.1.4.1	B	Measures quantities in the real world and uses the measures to solve problems
45	MA.C.3.4.1	H	Represents and applies geometric properties to solve real-world problems
46	MA.C.3.4.1	A	Represents and applies geometric properties to solve real-world problems
47	MA.B.2.4.1	H	Selects and uses appropriate measurements
48	MA.B.2.4.2	B	Solves real-world problems involving rated measures
49	MA.E.2.4.2	G	Determine probability for events
50	MA.E.3.4.2	D	Explains limitations of using statistical techniques and data in making inferences

Mathematics Assessment Two: Answer Key

1 The following table lists the masses of four of the moons of Jupiter.

Moon of Jupiter	Mass (in kg)
Callisto	1.08×10^{23}
Europa	4.80×10^{22}
Ganymede	1.48×10^{23}
Io	8.93×10^{22}

Which moon has the largest mass?
Analysis: *Choice C is correct.* The table gives mass in scientific notation, in which a number is expressed as a decimal multiplied by a power of ten. When 10 is raised to the nth power, it is multiplied by itself n times, and multiplying a number by 10^n is the same as moving the decimal point n places to the right (if n is positive) or n places to the left (if n is negative). You can determine the largest number by first comparing the powers of ten and then comparing the decimals. The masses of Callisto and Ganymede are larger than those of Europa and Io since 10^{23} is larger than 10^{22}. The mass of Ganymede is larger than that of Callisto since 1.48 is larger than 1.08.

2 Which of the choices below shows the numbers ordered from **largest** to **smallest**?
Analysis: *Choice F is correct.* Since $\sqrt{3} \approx 1.73$ and is smaller than 3, you can eliminate Choices G and I, in which $\sqrt{3}$ is shown as larger than 3. In general, $\sqrt{n}$ is smaller than n whenever n is greater than 1. Since $\frac{1}{\sqrt{3}} \approx \frac{1}{1.73}$ and is larger than $\frac{1}{3}$, you can eliminate Choice H, in which $\frac{1}{3}$ is shown as larger than $\frac{1}{\sqrt{3}}$. In general, increasing the denominator of a fraction makes the fraction smaller (if the whole is divided into more pieces, then each piece is smaller). Choice F shows the numbers listed in order from largest to smallest.

3 Which of the choices below shows the same number represented in three equivalent forms?
Analysis: *Choice D is correct.* Dividing 1 by 25 results in 0.04, so the first two terms are equivalent. To express a decimal as a percent, multiply the decimal by 100. Multiplying 0.04 by 100 yields 4, or 4%, so therefore all three terms are equivalent. In Choice A, 0.25 is equivalent to 25%, but these are not equivalent to 1/25 (which equals 0.04, or 4%). In Choice B, 1/4 is equivalent to 0.25, but these are not equivalent to 2.5% (which is equivalent to 0.025 = 25/1,000 = 1/40). In Choice C, 4/10 is equivalent to 40%, but these are not equivalent to 0.04 (which equals 4/100, or 4%).

4 Which of the following numbers is NOT equivalent to $\frac{1}{(-5)^2}$?

Analysis: *Choice G is correct.* Since multiplying two negative numbers yields a positive result, $1/(-5)^2 = 1/25$. The correct response is the one that shows a number other than 1/25. Remember that a negative exponent does not mean a negative number. It means the reciprocal. 5^2 and 5^{-2} are reciprocals of each other since $5^2 = 25$ and $5^{-2} = 1/25$. Also remember that the sign of the number is unaffected by the sign of the exponent. Positive numbers raised to any power, whether positive, negative, even, or odd, remain positive. Negative numbers raised to odd powers remain negative, but turn positive if raised to even powers. Choice F is $1/5^2 = 1/25$. Choice H is $5^{-2} = 1/5^2 = 1/25$. Choice I is $(-5)^{-2} = 1/(-5)^2 = 1/25$. However, Choice G is $1/5^{-2} = 1/(1/5^2) = 5^2 = 25$.

Mathematics Assessment Two: Answer Key

5 Cheryl receives a monthly allowance of $25.00. She also has a part-time job that pays $6.50 per hour, but she must pay $32.50 each month to ride the bus between her home and her job. After subtracting off her transportation costs, Cheryl deposits half of her earnings into her savings account every month. If she works w hours during the month, the amount of money she deposits into her savings account, S, can be represented by the formula below:
$S = 1/2(25 + 6.5w − 32.5)$
How much money will Cheryl add to her savings in September if she works 36 hours?
Analysis: *The correct answer is $113.25. The equation can be evaluated as follows:*
$S = 1/2(25 + 6.5w − 32.5)$
where $w = 36$.
$= 1/2(25 + 6.5(36) − 32.5)$
$= 1/2(25 + 234 − 32.5)$
$= 1/2(259 − 32.5)$
$= 1/2(226.5)$
$= 113.25$
Cheryl will add $113.25 to her savings in September.

6 Mr. Ruiz bought three books during his local bookstore's summer sale. Two of the books originally cost $12.95 and the third book originally cost $9.95. All three books were marked 10% off. He also paid a 6% sales tax. How much money did Mr. Ruiz spend altogether? Round your answer to the nearest penny.
Analysis: *The correct answer is $34.20.* The total price of the three books before the discount and tax is $35.85 (12.95 + 12.95 + 9.95 = $35.85) Since the discount is 10%, the sale price will be (100 − 10)% = 90% of the original price. To calculate the sale price, you should multiply the original price by 0.90. (0.90 x 35.85 = 32.265) The total cost of the books before tax is $32.265. (You should wait until you have calculated the final answer before rounding.) To

calculate the amount of the sales tax, multiply 32.265 by 0.06. (32.265 x 0.06 = 1.9359) To determine the total cost, add the sale price of the books and the amount of the sales tax, then round the sum to the nearest hundredth (the nearest penny). (32.265 + 1.9359 = 34.2009 ≈ 34.20) Alternatively, you can calculate the total cost by multiplying the sale price of the books by 1.06. (32.265 x 1.06 = 34.2009 ≈ 34.20) The correct answer is $34.20.

7 The bar graph below shows the population of each of four metropolitan areas in Florida.

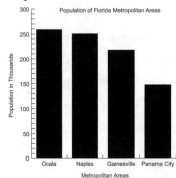

Which of the choices below is a good estimate of the total population of these four metropolitan areas?
Analysis: *Choice A is correct.* From looking at the bar graph, you could estimate the population of the four metropolitan areas as 260,000, 250,000, 220,000, and 150,000. The sum of these estimates is 880,000, which is close to 900,000.

8 Becky's school has arched doorways with the dimensions shown below.

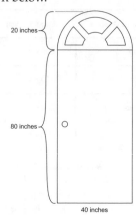

Which of the expressions below can be used to calculate the area of the doorway?
Analysis: *Choice G is correct.* From looking at the graph, you can see that the doorway can be divided into a rectangle on the bottom and a half circle on the top. The radius of the half circle is 20 inches (half the width of the doorway and the height of the top portion). The length and width of the rectangle are 80 inches and 40 inches. Therefore, the area of the doorway, A, is the sum of the area of a rectangle that measures 80 inches by 40 inches ($A_r = lw$) and one-half the area of a circle with a radius of 20 inches, ($A_c = \pi r^2$).

$A = A_r + A_c$ x $1/2$
$= $ length x width $+ \pi r^2$ x $1/2$
$= 80$ x $40 + \pi 20^2$ x $1/2$

This is the formula shown in Choice G. The formulas in Choices F and H calculate the perimeter of the rectangle instead of the area. Also, the formula in Choice H uses the diameter, not the radius, in the formula for the area of the circle. In the formula in Choice I, the area of the circle is not multiplied by 1/2.

Mathematics Assessment Two: Answer Key

9 What is the measure in inches of the curved edge of one section of the dartboard shown below? Round your answer to the nearest **hundredth**.

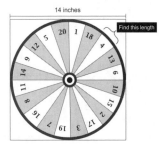

Analysis: *The correct answer is 2.20.* The dartboard is divided into 20 equal sections, so the curved edge of one section is 1/20th of the perimeter of the entire dartboard. Since the dartboard is a circle with a diameter of 14 inches, the perimeter of the dartboard can be found using the formula for the circumference of a circle: $C = \pi d = \pi \times 14 \approx 3.14 \times 14 = 43.96$ inches. Therefore, the curved edge of one section is $43.96 \times 1/20 = 43.96 \div 20 = 2.198 \approx 2.20$ inches.

10 A map shows the distance between Jacksonville and Orlando as 7.0 centimeters and the distance between Orlando and Fort Lauderdale as 10.6 centimeters.

If the actual distance between Jacksonville and Orlando is 140.4 miles, what is the actual distance between Orlando and Fort Lauderdale? Round your answer to the nearest **tenth** of a **mile**.

Analysis: *The correct answer is 212.6 miles.* Since the map is drawn to scale, the ratio of the distance shown on the map to the actual distance is always the same. Therefore, the ratio of the distance shown on the map between Jacksonville and Orlando to the actual distance between Jacksonville and Orlando is the same as the ratio of the distance shown on the map between Orlando and Fort Lauderdale to the actual distance between Orlando and Fort Lauderdale. To find the actual distance between Orlando and Fort Lauderdale, d, you can solve the following proportion:

$$\frac{7.0}{140.4} = \frac{10.6}{d}$$

$7.0d = 10.6 \times 140.4$

$7.0d = 1,488.24$

$d = 1,488.24 \div 7.0$

$\quad = 212.60571\ldots$

$\quad \approx 212.6$ miles

11 Linda wants to fill this rectangular sandbox two-thirds full of sand.

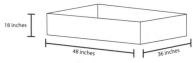

Linda can buy sand in bags that hold 1.5 cubic feet each. How many bags of sand will she need?

Analysis: *Choice B is correct.* Since the bags of sand Linda is buying are in cubic feet, we will need to get the volume of the sandbox in cubic feet. This will be easier if we convert its dimensions to feet first. Since there are 12 inches in a foot divide all of the boxes dimensions by 12 to get: 36 inches = 3 feet, 48 inches = 4 feet, and 18 inches = 1.5 feet. Since the sandbox is a rectangular prism, the volume, V, is given by the formula length x width x height. $V =$ length x width x height $= 3 \times 4 \times 1.5 = 18$ ft³. Multiply this number by 2/3 to find the volume of sand needed to fill the sandbox two-thirds full: $18 \times 2/3 = 12$ ft³. Each bag of sand holds 1.5 ft³ of sand. To find the number of bags Linda will need, divide the 12 cubic feet of sand required by 1.5. $12 \div 1.5 = 8$. Linda needs 8 bags of sand to fill the sandbox two-thirds full of sand.

12 Rhombus ABCD is shown in the graph below.

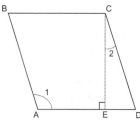

If ∠1 measures 127°, what is the degree measure of ∠2?

Analysis: *The correct answer is 37°.* In all parallelograms (and a rhombus is a type of parallelogram), opposite sides are parallel and opposite angles are equal. This means that segment BC is parallel to segment AD and ∠1 is equal to ∠BCD. So $m\angle BCD = 127°$. Since CE is perpendicular to AD, it must also be perpendicular to any line that is parallel to AD, such as BC. If CE is perpendicular to BC then ∠BCE is a right angle and ∠2 is equal to $127° - 90° = 37°$.

Mathematics Assessment Two: Answer Key

13 Parallelogram LMNO is shown below.

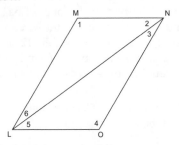

Which of the following equations is NOT true for the measures of angles 1, 2, 3, 4, 5, and 6?
Analysis: *Choice H is correct.* The table below summarizes what we know about these angles and how it is known.

What is known	How it is known
$m\angle 1 = m\angle 4$	Opposite angles of a pallellogram are equal
$m\angle 2 = m\angle 5$	MN is parallel to LO with LN as a transversal; opposite interior angles are equal
$m\angle 3 = m\angle 6$	LM is parallel to ON with LN as a transversal; opposite interior angles are equal

Choice H's equation is not true unless it can be stated that $m\angle 5$ is equal to $m\angle 6$ and this could only be true if LMNO is a rhombus, which is not given. To see why this statement is untrue, temporarily assume that $\angle 5$ and $\angle 6$ are equal. If $\angle 5$ and $\angle 6$ are equal, it also means that $\angle 5$ and $\angle 3$ are equal. If $\angle 5$ and $\angle 3$ are equal, then triangle LNO is isosceles and segment LO equals segment NO. But segment LO does not equal segment NO unless LMNO is a rhombus, which is not given.

14 Triangle QRS is inscribed in a circle centered at point P as shown in the sketch below. The radius of circle P is 3 centimeters and segment SP is perpendicular to segment QR.

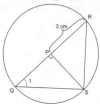

What is the degree measure of $\angle 1$?
Analysis: *The correct answer is 45°.* Since all radii of a circle are equal, segment QP and segment PS are equal and triangle QPS is an isosceles triangle. This means that the opposite angles, $\angle 1$ and $\angle PSQ$, are also equal and together they add up to 90° since the third angle in triangle QPS is right angle $\angle QPS$. (180° – 90° = 90°). If two angles are equal and the sum of their measures is 90°, then each angle must measure 45°. Therefore, $\angle 1$ measures 45°.

15 Triangle QRS is inscribed in a circle centered at point P as shown in the sketch below. The radius of circle P is 3 centimeters and segment SP is perpendicular to segment QR.

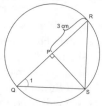

What is the length in centimeters of segment QS? Express your answer as a decimal rounded to the nearest **hundredth**.
Analysis: *The correct answer is 4.24 cm.* Since all radii of a circle are equal, segment RP, segment QP, and segment PS are all equal to 3 cm and triangle QPS is an isosceles right triangle. (It's given that segment SP is perpendicular to segment QR.) This means segment QS is the hypotenuse of a right triangle and its

length can be found using the Pythagorean theorem.
$$a^2 + b^2 = c^2; 3^2 + 3^2 = c^2; 9 + 9 = c^2;$$
$$c^2 = 18; c = \sqrt{18}; c = \sqrt{9 \times 2};$$
$$c = \sqrt{9} \times \sqrt{2}; c = 3\sqrt{2} \approx 4.24$$

16 Triangle ABC is shown below. Suppose there is a triangle DEF that is similar to triangle ABC. The shortest side of triangle DEF is segment DF, which measures 10 cm.

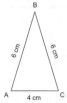

What is the perimeter of triangle DEF in centimeters?
Analysis: *The correct answer is 40 cm.* Since triangles ABC and DEF are similar, their corresponding sides and perimeters are proportional. The length of the shortest side of DEF, segment DF, is 10 cm. The shortest side of ABC, segment AC, is 4 cm. DF corresponds to AC in a ratio of 10:4. Therefore, the other two pairs of corresponding sides are also in a ratio of 10:4. Perimeter is just the sum of the sides, so the perimeters of these two triangles also have the ratio of 10:4. The perimeter of triangle ABC is 16 cm (6 + 6 + 4 = 16) This is a proportion problem. Proportions are usually set up with one fraction being set equal to another fraction. Set up a proportion and solve for the unknown perimeter of triangle DEF.

$$\frac{10}{4} = \frac{x}{16}$$

$$4x = 16 \times 10$$

$$4x = 160$$

$$x = 40 \text{ cm}$$

Mathematics Assessment Two: Answer Key

17 ABC and DEF are similar right triangles and segment AC corresponds to segment DF as shown below.

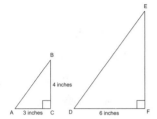

Which of the following statements is NOT true?

Analysis: *Choice B is correct.* Similar triangles are the same shape, but not necessarily the same size. If triangles are similar, their corresponding angles are equal and their corresponding sides are proportional. That is, each pair of corresponding sides has the same ratio. The table below summarizes and compares the properties of these two similar triangles.

ΔABC	ΔDEF
AC = 3 inches	DF = 6 inches
BC = 4 inches	EF = 8 inches
AB = 5 inches	DE = 10 inches
Perimeter = 12 in	Perimeter = 24 in
Area = 6 in²	Area = 24 in²
∠A = ∠D ∠B = ∠E ∠C = ∠F	∠A = ∠D ∠B = ∠E ∠C = ∠F

Since segment AC corresponds to segment DF and their lengths are 3 inches and 6 inches, then corresponding sides are in the ratio 3:6 = 1:2. Therefore, the other two pairs of corresponding sides are also in a ratio of 1:2. Perimeter is just the sum of the sides, so the perimeters of these two triangles also have the ratio of 1:2. You can calculate the length of segment AB using the Pythagorean theorem:

$a^2 + b^2 = c^2$; $3^2 + 4^2 = c^2$; $9 + 16 = c^2$; $c^2 = 25$; $c = 5$. So ΔABC's sides are 3

in, 4 inches, and 5 inches. Since the ratio between the corresponding sides of these triangles is 1:2, , all we have to do to get ΔDEF's sides is to double ΔABC's sides. ΔDEF's sides are 6 inches, 8 inches, and 10 inches. If the ratio had been harder to calculate, all we'd have to do is set up and solve some proportions. ΔABC's perimeter is 12 inches (3 + 4 + 5 =12 inches). ΔDEF's perimeter is 24 inches (6 + 8 + 10 = 24 inches). ΔABC's area is 6 in² (1/2 x 3 x 4 = 6 in²). ΔDEF's area is 24 in² (1/2 x 6 x 8 = 24 in²) Therefore, the statement in Choice B is NOT true. Remember that the areas of two similar triangles are NOT in the same ratio as their sides.

Important Test Hint: Both of the triangles in this problem are special. Right triangles like these that have all three sides as integers are called Pythagorean Triples and are much rarer and much more popular with test-makers than most right triangles. Pythagorean Triples come in "families." If you take any Pythagorean Triple and multiply each of its sides by the same whole number, you will get another Pythagorean Triple in the same family. ΔABC is one of the "most famous" of all right triangles and it's the smallest Pythagorean Triple. It's called the 3-4-5 triangle, after the length of its sides. Other members of this family are: the 6-8-10 (multiplied by 2), the 9-12-15 (multiplied by 3), the 12-16-20 (multiplied by 4), the 15-20-25 (multiplied by 5), and so on. All of these triangles are really the 3-4-5 triangle in disguise. When you get a problem involving a Pythagorean Triple, it's great, because you don't need the Pythagorean Theorem to find the missing side. For example, if the given legs are 3 and 4, you know the hypotenuse must be 5. If the given hypotenuse is 25 and the given leg is

15, the other leg must be 20. It will save a lot of test time if you memorize the 3-4-5 triangle and how to generate other members of this family. The only other Pythagorean Triple commonly used by test-makers is the 5-12-13 triangle. It is worth remembering this one also. ($a^2 + b^2 = c^2$; $5^2 + 12^2 = 13^2$; 25 + 144 = 169)

18 Quadrilateral JKLM is shown in the graph below.

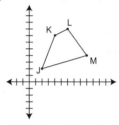

Suppose JKLM is reflected once over the *x*-axis and then translated by -2 along the *y*-axis. What are the new coordinates of Point M?

Analysis: *Choice F is correct.* The coordinates of point M (9, 4). Reflecting JKLM once over the *x*-axis and then translating JKLM -2 along the *y*-axis is the same as multiplying each of the y-coordinates by -1 and then subtracting 2 from each *y*-coordinate. This is shown in Choice F. If trying to figure out which numerical operations to perform on which coordinates confuses you, try solving this problem graphically. When reflecting an object across a line, each point in the reflected image must be the same distance from the line of reflection as the original. In this case the line of reflection is the *x*-axis, so each point in the reflected quadrilateral must be the same distance from the *x*-axis as the original quadrilateral. Notice that in the original quadrilateral, Point M is at (9, 4). When Point M is reflected across the *x*-axis, its *x*-coordinate will not change. It will still be 9. However, to make the reflection of Point M 4 units from the *x*-axis, its *y*-

Mathematics Assessment Two: Answer Key

coordinate will have to change from positive to negative. The y-coordinate must be -4. The same thing happens to all the other points. When the point is translated by -2 along the y-axis, all the points move down two units.

19 JKL is a triangle with the coordinates shown below.

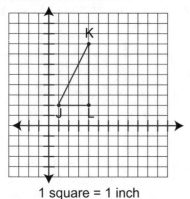

1 square = 1 inch

What is the length of segment JK in inches? Round your answer to the nearest **hundredth**.

Analysis: *The correct answer is 6.71 inches.* The coordinates of points J, K, and L are J(1, 2), K(4, 8), and L(4, 2). You can see from the graph that JKL is a right triangle with hypotenuse segment JK. By counting squares you can find the length of the triangle's legs; segment JL = 3 and segment KL = 6. Apply the Pythagorean Theorem to find the hypotenuse: $a^2 + b^2 = c^2$; $3^2 + 6^2 = c^2$; $9 + 36 = c^2$; $c^2 = 45$; ≈ 6.71 in^2.

Alternatively, you can find the length of segment JK using the coordinates of Points J and K with the Distance formula.

20 The graph of line ℓ is shown below.

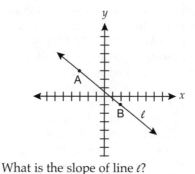

What is the slope of line ℓ?

Analysis: *Choice A is correct.* You can determine the slope of line ℓ using the two points shown on the graph. The slope of a line is its rise over its run (*rise/run*) or the change in y of the change in x ($\Delta y / \Delta x$). To get from Point A to Point B, you must go down 4 units and right 5 units. This is a change in y of -4 and change in x of $+5$. Therefore, the slope is $-4/5 = -0.8$. The slope can also be found using the formula slope = *rise/run* = $(y_2 - y_1)/(x_2 - x_1)$, where (x_1, y_1) = A(-3, 3) and (x_2, y_2) = B(2, -1).

slope = $(-1 - 3)/(2 - (-3))$
     = $-4/2 + 3$
     = $-4/5$
     = -0.8

A line has negative slope when the value of y decreases as x increases. Choices C and D show a positive slope, which is incorrect. Choice B results from calculating *run/rise* = $5/-4$ instead of *rise/run* = $-4/5$ and is incorrect.

21 The swimming pool at Northland Swim Club is a parallelogram and is drawn to scale in the graph below.

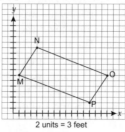

2 units = 3 feet

What is the perimeter of the actual swimming pool in feet?

Analysis: *Choice I is correct.* Your first impulse may be to use the distance formula, D = $\sqrt{(x_2 - x_1)^2 + (y_2 - y_1)^2}$ which is really a transformation of the Pythagorean Theorem. Neither of these is actually required in this case. Whenever you need to find the distance between two points on a coordinate plane, first check to see if any Pythagorean Triples are being used. In this case two different Pythagorean Triples are imbedded into the solution sketch. The shorter sides of the pool are completed by 3-4-5 triangles. The longer sides of the pool are completed by 5-12-13 triangles. (See Important Test Hint page 217.) These are favorite right triangles of test-makers. Now that the length of the pools sides is known, just add them to find the perimeter in squares (5 + 13 + 5 + 13 = 36 squares). Since 2 squares equals 3 feet, set up the following proportion and solve: $2/3 = 36/x$; $2x = 108$; $x = 54$ feet. Alternatively, you could use the distance formula to find the length of the pool sides. The coordinates of points M, N, O, and P are M(1, 7), N(4, 11), O(16, 6), and P(13, 2).

MN = $\sqrt{(4 - 1)^2 + (11 - 7)^2}$ where M(x_1, y_1) = (1, 7) and N(x_2, y_2) = (4, 11)

Mathematics Assessment Two: Answer Key

$$= \sqrt{3^2 + 4^2}$$
$$= \sqrt{9 + 16}$$
$$= \sqrt{25}$$
$$= 5$$

$NO = \sqrt{(16 - 4)^2 + (6 - 11)^2}$ where
$N(x_1, y_1) = (4, 11)$ and $O(x_2, y_2) = (16, 6)$

$$= \sqrt{12^2 + (-5)^2}$$
$$= \sqrt{144 + 25}$$
$$= \sqrt{169}$$
$$= 13$$

Since it is given that MNOP is a parallelogram, then segment PO = segment MN and segment MP = segment NO. Therefore, the perimeter of the pool in graph squares is $5 + 13 + 5 + 13 = 36$ squares. To calculate the perimeter of the actual pool in feet, you must use the fact that 2 squares = 3 ft. Divide the perimeter in graph squares by 2 and then multiply by 3.

$(36 \div 2) \times 3 = 18 \times 3 = 54$ ft

22 The function table below shows values of x and y that satisfy the equation $-2x^3 + y = 0$.

x	y
0	0
1	2
2	16
3	54
4	128
5	

What is the value of y when $x = 5$?
Analysis: *The correct answer is 250.*
The equation $-2x^3 + y = 0$ is equivalent to $y = 2x^3$. Substituting 5 for x in the equation,
$y = 2(5)^3$
   $= 2(5 \times 5 \times 5)$
   $= 2(25 \times 5)$
   $= 2(125)$
   $= 250$

23 For the function table shown below, what is the value of y when $x = 5$?

x	y
0	1
1	4
2	13
3	28
4	49
5	
6	109

Analysis: *Choice C is correct.* Since no equation is given, you will need to look for a pattern in the y values given in the table. You can continue that pattern to find the value of y when $x = 5$.

Method 1			
x	y	y–(previous y)	relationship
0	1		
1	4	4 – 1 = 3	3 = 3 x 1
2	13	13 – 4 = 9	9 = 3 x 3
3	28	28 – 13 = 15	15 = 3 x 5
4	49	49 – 28 = 21	21 = 3 x 7
5	76		
6	109		

Each difference is 3 x the next odd number

This must be 3 x 9 = 27. So 49 + 27 = 76. When $x = 5$, $y = 76$

If 76 is correct for $x = 5$, then it must produce 109 here. This must be 3 x 11= 33. So 76 + 33 = 109. So, 76 is correct.

Alternatively, you could notice that the difference between the y values increases by 6 each time.

Method 2			
x	y	How to calculate this y-value	What to add to this y-value to get the next y-value
0	1	given	3 + 0 x 6 = 3
1	4	1 + 3 = 4	3 + 1 x 6 = 9
2	13	4 + 9 = 13	3 + 2 x 6 = 15
3	28	13 + 15 = 28	3 + 3 x 6 = 21
4	49	28 + 21 = 49	3 + 4 x 6 = 27
5	76	49 + 27 = 76	3 + 5 x 6 = 33
6	109	76 + 33 = 109	3 + 6 x 6 = 39

Mathematics Assessment Two: Answer Key

24 The Turners pay a monthly fee of $34.99 for their phone service. For this monthly fee, they can make 300 minutes' worth of phone calls at no additional charge. For every minute over 300, they pay an additional $0.40. Assuming that the Turners use their phone service for more than 300 minutes each month, which of the following equations shows *P*, the Turners' total phone bill, as a function of *m*, the total number of minutes?

Analysis: *Choice F is correct.* The Turners' total phone bill is the monthly fee of $34.99 plus $0.40 times the number of minutes over 300, which is equal to the total number of minutes, *m*, minus 300. The total phone bill can be calculated with the following equation:

$$P = 34.99 + 0.40(m - 300)$$
$$= 34.99 + 0.40m - 120$$
$$= 0.40m - 85.01$$

This is the equation shown in Choice F, but remember that this equation only holds true if the total number of minutes, *m*, is over 300. The equation in Choice G is equivalent to $P = -34.99 + 0.40(m + 300)$ and is incorrect. The equation in Choice H is equivalent to $P = 34.99 + 0.40(m + 300)$ and is incorrect. The equation in Choice I is equivalent to $P = -34.99 + 0.40(m - 300)$ and is incorrect. The equations in Choices G and H calculate the number of minutes over 300 as $m + 300$ instead of $m - 300$. The equations in Choices G and I subtract the monthly fee of $34.99 instead of adding it.

25 Rikkhyia wants to calculate the cost of driving her car to work for one week. She works five days each week and the distance between her home and her workplace is 14.1 miles. Her car averages 30 miles per gallon of gasoline. In addition to paying for gas, she must also pay $4.00 per day for parking. Let *g* equal the cost of a gallon of gasoline. Which of the equations below shows Rikkhyia's total cost, *C*, as a function of *g*?

Analysis: *Choice B is correct.* To drive her car to work for one week, Rikkhyia will pay the cost of parking for 5 days ($4.00 x 5 = $20.00) plus the cost of gasoline for making 10 trips back and forth to her workplace. Since the distance between her home and her workplace is 14.1 miles, the total mileage of these 10 trips is 14.1 x 10 = 141. So, the cost of gasoline is the price per gallon of gasoline, *g*, times the number of gallons of gasoline needed to drive 141 miles. The number of gallons of gasoline is 141 divided by the car's gas mileage in miles per gallon, or 141/30. Therefore, the following equation shows Rikkhyia's total cost of driving her car to work for one week: $C = 141/30$ x $g + 20$

This is the equation shown in Choice B. The equations in Choices C and D calculate the cost of parking for one day ($4.00) instead of one week ($20.00) and is incorrect. The equations in Choices A and D calculate the number of gallons of gasoline incorrectly, dividing gas mileage by miles driven (30/141) instead of dividing miles driven by gas mileage (141/30).

26 Mary Ann is preparing to move and is packing her clothes into boxes that measure 12 inches by 15 1/2 inches by 10 inches. She wants to find new boxes that will allow her pack four times the volume of clothing in each box. Of the boxes described below, which one should Mary Ann NOT choose?

Analysis: *Choice G is correct.* Since the volume of a rectangular box is found by multiplying length by width by height, you can increase the volume by a factor of 4 by increasing one of the dimensions by a factor of 4 or by increasing two of the dimensions by a factor of 2. This is shown in Choices F, H, and I. In Choice F, the depth is increased by a factor of 4 (10 x 4 = 40) and the length and width remain the same. In Choice H, the length and depth are both increased by a factor of 2 (12 x 2 = 24 and 10 x 2 = 20) and the width remains the same. In Choice I, the width and depth are both increased by a factor of 2 (15-1/2 x 2 = 31 and 10 x 2 = 20) and the length remains the same. In Choice G, all three dimensions are increased by a factor of 2 (12 x 2 = 24, 15-1/2 x 2 = 31, and 10 x 2 = 20), which increases the volume by a factor of 8, not 4, since 24 x 31 x 20 = (2 x 12) x (2 x 15-1/2) x (2 x 10) = (2 x 2 x 2) x (12 x 15-1/2 x 10) = 8 x (12 x 15-1/2 x 10).

Mathematics Assessment Two: Answer Key

27 Janice makes lemonade using lemons, sugar, and water. She uses the juice from 4 lemons and 3 cups of sugar for every gallon of lemonade. If she used 18 lemons altogether, how many cups of sugar did she use? Express your answer as a fraction or a decimal.
Analysis: *The correct answer is* $13\ 1/2 = 13.5$. Let c equal the number of cups of sugar Janice used. Since the number of lemons and the number of cups of sugar are in the ratio 4:3, you can solve the following proportion to determine c:

$$\frac{4}{3} = \frac{18}{c}$$
$$4c = 18 \times 3$$
$$4c = 54$$
$$c = \frac{54}{4}$$
$$= 13\frac{2}{4} = 13\frac{1}{2} = 13.5$$

28 Karen runs for 45 minutes every morning. She averages 5.5 miles a day. If she wants to run an average of 8 miles each day, by how much does she need to increase her average speed?
Analysis: *Choice D is correct.* To find Karen's current speed, use the formula $d = rt$ (distance = rate $\times$ time). Since you're looking for an answer expressed in miles per hour, you should first convert the time given in the problem from minutes to hours: 45 minutes = $45/60$ hour = 0.75 hour. Since Karen is currently running 5.5 miles in 0.75 hour, you can solve the following equation for r: $5.5 = r \times 0.75$; $r = 5.5/0.75 \approx 7.33$ miles/hour. Now find the speed Karen will need to have in order to run 8 miles in 0.75 hour: $8 = r \times 0.75$; $r = 8/0.75 \approx 10.67$ miles/hour. The amount by which Karen will need to increase her speed is the difference between this speed, 10.67 miles/hour, and her current speed, 7.33 miles/hour. Therefore, in order to run 8 miles in 0.75 hour,

Karen will need to increase her speed by $10.67 - 7.33 = 3.34$ miles/hour. Since 3.34 is approximately 3 1/3, the correct answer is Choice D.

29 The equation $y = 2x + 2$ is graphed below.

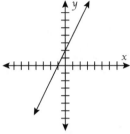

Which of the following graphs shows this equation with the first 2 replaced by a 3?
Analysis: *Choice F is correct.* This is the graph of $y = 3x + 2$.

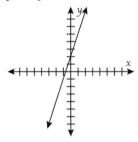

Replacing the first 2 in the equation with a 3 changes the slope of the line from 2 to 3 (meaning that when x increases by 1 unit, then y increases by 3 units instead of 2); the y-intercept (the value of y when $x = 0$) does not change. The graph of $y = 2x + 3$ is shown in Choice G. This graph shows a line with the same slope as the original equation, but a y-intercept of 3 instead of 2 (meaning that the line has been shifted up 1 unit). The graph of $y = x + 2$ is shown in Choice H. This graph shows a line with the same y-intercept as the original, but a slope of 1 instead of 2 (meaning that when x increases by 1, then y increases by 1 instead of 2). The graph of $y = 3x + 3$ is shown in Choice I. This graph shows a line with a slope of 3 instead of 2, but also a y-intercept of 3 instead of 2 (meaning that the line is also shifted up 1 unit).

30 Alex and John are both walking to school. They have agreed to meet on the school grounds as soon as they arrive. Alex's home is 0.6 mile from the school and he walks an average of 4 miles per hour. John's home is 0.9 mile away and he walks an average of 4.5 miles per hour. If they leave home at the same time, how many minutes will one of the boys wait for the other?

Analysis: The correct answer is 3 minutes. To find out how long it takes each boy to walk to school, use the formula $d = rt$, distance = rate $\times$ time. To determine Alex's time, t_A, solve the following equation where $d_A = 0.6$ and $r_A = 4$: $d = rt$; $0.6 = 4 \times t_A$; $t_A = 0.6/4 = 0.15$ hour. To determine John's time, t_J, solve the following equation where $d_J = 0.9$ and $r_J = 4.5$: $d = rt$; $0.9 = 4.5 \times t_J$; $t_J = 0.9/4.5 = 0.2$ hour. Since John's time is greater than Alex's time, it will take John longer to reach the school and Alex will have to wait for him there. To determine how many minutes Alex will wait, subtract Alex's time from John's time, then multiply the result by 60 to convert from hours to minutes: $t_J - t_A = 0.2 - 0.15 = 0.05$ hour; $0.05 \times 60 = 3$ minutes. Therefore, Alex will wait 3 minutes for John. Alternatively, you could use ratios to determine the number of minutes it took each boy to walk to school.

$$\frac{0.6}{4} = \frac{A}{60} \qquad \frac{0.9}{4.5} = \frac{J}{60}$$
$$4A = 36 \qquad\qquad 4.5J = 54$$
$$A = 9 \qquad\qquad\quad J = 12$$

Since $12 - 9 = 3$, Alex will wait 3 minutes.

Mathematics Assessment Two: Answer Key

31 Isaac is going to the state fair. He can either pay $12.00 for admission plus $1.75 for every ride, or he can pay $30.00 for admission and ride for free all day. What is the maximum number of rides Isaac can go on and still pay less with the $12.00 admission price?

Analysis: *Choice B is correct.* If Isaac pays $12.00 for admission, then his total cost can be calculated with the expression $12 + 1.75r$, where r is the number of rides. If Isaac's total cost with the $12.00 admission price is less than his total cost with the $30.00 admission price, then the following expression is true: $12 + 1.75r < 30$. The largest number of rides Isaac can go on and still pay less with the $12.00 admission price is the largest whole-number value of r for which this expression is true. You can solve this expression for r as follows: $12 + 1.75r < 30$; $1.75r < 30 - 12$; $1.75r < 18$; $r < 18/1.75$; $r < 10.285714... \approx 10.29$. Therefore, Isaac can go on as many as 10 rides and still pay less with the $12.00 admission price.

32 Jonathan is painting 6 circular columns on the front porch of an apartment building. For each column, he can use the formula $S.A. = 2\pi rh$ to calculate the area of a column's exposed surface that he will need to paint. Neither the upper nor the lower base of the column are visible, so they will not be painted. The height of each column is 9.5 feet and the diameter is 3 feet. The paint he is using will cover 350 square feet per gallon, but it is sold by the quart. How many quarts of paint should he buy?

Analysis: *The correct answer is 7 quarts.* Use the formula to calculate the area of a column's visible surface. If the diameter of a column is 3 ft, then the radius is 1.5 ft. $S.A. = 2\pi rh$; $S.A. \approx 2(3.14)(1.5)(9.5) \approx 89.49$ ft^2. There are 6 columns, so he needs to paint $6 \times 89.49 = 536.94$ ft^2. Since a gallon of paint will cover 350 ft^2,

you can determine how many gallons of paint Jonathan will need by dividing the total surface area by 350: $536.94 \div 350 \approx 1.53$ gallons. There are 4 quarts in a gallon so Jonathan will need $1.53 \times 4 \approx 6.12$ quarts. He can't buy part of a quart, so he should buy 7.

33 The Argosy Bookshop has the following inventory:

Types of Books	Number of Books
Fiction & Poetry	9,042
History & Biography	10,118
Travel	5,431
Other Nonfiction	4,087

At the end of a week, they have sold 270 books in the fiction and poetry category. If they've sold the same percentage of travel books, how many travel books have they sold? Round your answer to the nearest **whole number**.

Analysis: *The correct answer is 162 books.* Let b_t equal the number of travel books sold. Since they sold the same percentage of travel books and fiction and poetry books, you can calculate b_t by solving the following proportion: $270/9,042 = b_t/5,431$; $b_t \times 9,042 = 270 \times 5,431$; $b_t \times 9,042 = 1,466,370$; $b_t = 1,466,370/9,042 \approx 162.17319...$ Rounded to the nearest whole number, the number of travel books sold was 162.

34 Kevin was interested in the different combinations of pets owned by his classmates. He conducted a survey in which he asked each student whether or not they owned a dog, a cat, or a fish. He represented his findings with the Venn diagram shown below, in which Circle A is the set of all dog owners, Circle B is the set of all cat owners, and Circle C is the set of all fish owners.

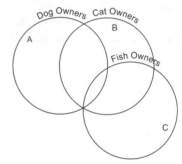

Which of the following statements is true for the students in Kevin's class?

Analysis: *Choice H is correct.* Choice H is true since Circle A, the set of all dog owners, and Circle C, the set of all fish owners, do not overlap in the diagram except where they also overlap with Circle B, the set of all cat owners. Therefore, everyone who owns both a dog and a fish also owns a cat. Students in Area E, own a dog, a cat, and a fish, so Choice F is incorrect. Students in Area D own a dog and a cat, but not a fish, so Choice G is incorrect. Students in Area F own a cat and a fish, but not a dog, so Choice I is incorrect.

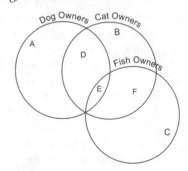

Mathematics Assessment Two: Answer Key

35 The heights of 14 people are shown in the line plot below.

What is the median height for this group of people? Express your answer in **inches**.

Analysis: *The correct answer is 64.5.* The median of a set of numbers is found by arranging the numbers in order and then selecting the middle number or, if there are an even number of numbers, determining the mean of the two middle numbers. Since there are 14 numbers in this set of heights, the median is the mean of the numbers that are 7th and 8th in order. You can count from either end of the line plot to find the heights that are 7th and 8th in order: 64 and 65. The mean of these two numbers is $(64 + 65)/2 = 129/2 = 64.5$ in.

36 Suppose you were told that the mean of the set of all test scores for a class of 28 students is 76. Which of the followings statements is true?

Analysis: *Choice A is correct.* The mean is calculated by adding all the test scores and then dividing the result by the number of students in the class, 28. If you multiply 76 by 28, the result will be the total of all the students' test scores. The statement in Choice B is correct only if 76 is also the median of the set (the middle value, or the mean of the two middle values, when the test scores are arranged in order). Choice C is correct only if 76 is also a mode of the set (the value that occurs most often in the set of all test scores, or one such value). Choice D is incorrect because it applies to the median of a set of scores, not the mean which is not necessarily the middle value in the set. The range of the set is the difference between the highest and the lowest test scores. If m is the median of the set, then the following is a true statement: All of the test scores are between the numbers $m - r/2$ and $m + r/2$.

37 Jay has signed up to play baseball at his neighborhood recreation center. He will be assigned to one of the following teams: Marlins, Astros, Diamondbacks, or Dodgers. The coach will assign him to play one of the following positions on the team: first baseman, second baseman, third baseman, pitcher, catcher, shortstop, right fielder, center fielder, or left fielder.

Teams		Positions
Marlins		first baseman
Astros		second baseman
Diamondbacks		third baseman
Dodgers		pitcher
		catcher
		shortstop
		right fielder
		center fielder
		left fielder

If the assignments are made at random, what is the probability that Jay will be a fielder for the Marlins? Express your answer as a fraction.

Analysis: *The correct answer is 1/12.* The probability that Jay will be assigned to any one of the teams is 1 over the total number of teams, which is 4. So the probability that Jay will be selected for the Marlins is 1/4. Since there are 3 fielding positions out of a total of 9 positions, the probability that Jay will be assigned to a fielding position is $3/9 = 1/3$. Since the team assignment and the position assignment are independent of each other, you can find the probably that Jay will be assigned to the Marlins AND assigned to be a fielder by multiplying the probabilities of these two events. Therefore, the answer is $1/4 \times 1/3 = 1/12$.

38 Suppose you are playing a board game that uses two spinners and a die to determine a player's next move. A player spins both arrows and rolls the die, and then moves ahead by the number of spaces shown on the die only if both arrows land on black. What are the chances that you will move ahead on your next turn?

Analysis: *Choice G is correct.* Since the spinners are divided into equal parts, the probability that the arrow will land on black equals the number of black parts over the total number of parts. For the spinner on the left, the probability that the arrow will land on black is $4/12 = 1/3$. For the spinner on the right, the probability that the arrow will land on black is $5/10 = 1/2$. Since the outcomes of the two spins are independent, you can find the probability that both outcomes will be successful by multiplying. Therefore, the probability that both arrows will land on black is $1/3 \times 1/2 = 1/6 \approx 0.167$.

Mathematics Assessment Two: Answer Key

39 Every student at Fort Hayes High School takes one, and only one, foreign language class. The following table shows the number of students who study each of the languages offered at Fort Hayes.

	Spanish	French	Latin	Japanese
9th Grade	127	62	27	13
10th Grade	102	71	33	25
11th Grade	82	79	45	38
12th Grade	90	57	75	40

What is the probability that an 11th grader chosen at random is taking French? Express your answer as a percent rounded to the nearest whole number.
Analysis: *The correct answer is 32.* The probability that an 11th grader chosen at random is taking French is the number of 11th graders taking French over the total number of 11th graders. The number of 11th graders taking French is 79 and the total number of 11th graders is $82 + 79 + 45 + 38 = 244$. Therefore, the probability that a randomly chosen 11th grader is taking French is $79/244 = 0.3237704\ldots \approx 0.32$, or 32%.

40 Consolidated Fish Company has a pier-side location where it buys shrimp directly from the fishing fleet. It pays $2.25 per pound for shrimp. Usually the shrimp are brought to market in 40-pound capacity plastic containers. Which of the following represents a formula that could be used for buying any quantity of shrimp, where P is the overall amount paid for a catch and S is the weight of the shrimp being sold?
Analysis: *Choice B is correct.* The equation in Choice B correctly multiplies the amount of shrimp (the unknown, which is represented by the variable S) by the price per pound ($2.25), which yields the price paid (represented by the variable P). It doesn't matter what size the container is when bringing the fish to market. The number 40 is extraneous information.

41 The process of simplifying the equation $2y = 2x + 2$ best demonstrates which of the following properties?
Analysis: *Choice H is correct.* The distributive property maintains that $a(b + c) = ab + ca$.

42 In mathematics, what is a dilation?
Analysis: *Choice B is correct.* A dilation is a reduced or an enlarged representation of a figure, shown in proper proportion.

43 Jane is preparing to mail an important set of design plans. She has rolled the plans up and placed them inside a cylinder-shaped tube that is 27 inches tall and has a radius of 2 inches. Jane plans to select a rectangular box that will allow at least 2 inches of packing material on all sides of the plans. What is the smallest box she can use for mailing the plans?
Analysis: *Choice H is correct.* Since the plans are 27 inches tall, 2 inches of clearance on the top and on the bottom requires a box at least 31 inches tall. If the radius is 2 inches, then the diameter is 4 inches. Two-inch clearance requires width and depth of 8 inches. So the box would need to be 8 inches by 8 inches by 31 inches.

44 Mason dug a trench 60 feet long, 2 feet wide, and 2 feet deep as part of a construction project. This soil will have to be removed in pickup truck loads until it has been completely cleared from the property. The pickup truck bed will safely carry a load of soil that measures 4 feet by 8 feet by 2 feet. How many truckloads will have to be made to clear the soil from the property?
Analysis: *Choice B is correct.* There are 240 cubic feet of soil to be removed ($60 \times 2 \times 2$), and the truck bed will remove it at 64 cubic feet per trip ($4 \times 8 \times 2$), or 3.75 loads ($240 \div 64 = 3.75$). Since the truck can't make part of a trip, the answer must be 4 because 3 trips will not be enough to do the job.

45 A large rectangular-shaped field measures 600 feet by 800 feet. How long would a fence need to be if it were installed from one corner of this field to the opposite diagonal corner?
Analysis: *Choice H is correct.* The field is actually a Pythagorean Triple, the 3-4-5 right triangle with each side multiplied by 200. Since the triangles are similar, their corresponding sides are in the same proportion: $3 \times 200 = 600$, $4 \times 200 = 800$, $5 \times 200 = 1000$. If you did not notice that the field is a multiple of the 3-4-5 right triangle, you could still apply the Pythagorean Theorem: $a^2 + b^2 = c^2$; $600^2 + 800^2 = c^2$; $360{,}000 + 640{,}000 = c^2$; $c^2 = 1{,}000{,}000$; $c = 1{,}000$.

Mathematics Assessment Two: Answer Key

46 A ladder that is eight feet tall has been leaned against a wall at an angle, and the base of the ladder is firmly secured at a point exactly three feet from the wall. Rounded to the nearest foot, how high up the wall is the top of the ladder?
Analysis: *Choice A is correct.* Use the Pythagorean Theorem:
$a^2 + b^2 = c^2$; $3^2 + b^2 = 8^2$; $9 + b^2 = 64$;
$b^2 = 64 - 9$; $b^2 = 55$; $b = \sqrt{55} \approx 7.4$.
An easier way to solve this problem is to notice that Choices B, C, and D could not work as answers since the hypotenuse itself is 8 feet and the unknown leg must be shorter.

47 Rene sees a box on a shelf and notices its height is about 12 inches and its width is about 10 inches. On the end of the box that she can see, it is labeled 3 cubic feet of storage, but the rest of the box extends back into the shelf and cannot be seen. About how long should the box be if 3 cubic feet is its actual volume?
Analysis: *Choice H is correct.* Convert all the measures to the same unit. Since the possible answers are given in inches, convert the volume of 3 cubic feet into cubic inches. A cubic foot is 1 ft x 1 ft x 1 ft. Each foot is the same as 12 inches, so 1 cubic foot = 12 in x 12 in x 12 in = 1,728 cubic inches. The volume of the box is 3 cubic feet so this is equivalent to 5,184 cubic inches (3 x 1,728 = 5,184). The volume of a box is found using the formula $V = lwh$. We know the volume, the height, and the width of the box. We want to find the length. Plug values into the formula for volume and solve for the unknown length: $V = lwh$; $5,184 = l$ x 10 x 12; $5,184 = l$ x 120; $l = 5,184 \div 120$; $l = 43.2$ inches.

48 The floor plan of a new office complex calls for 35,000 total square feet. Construction costs for this facility are estimated to be $122.00 per square foot, based on the design plans. How much money will this office complex cost to build if the estimate is correct?
Analysis: *Choice B is correct.* Multiply the total number of square feet by the cost per square foot of space.

49 What is the probability that a black marble in a bag with 3 blue marbles will be drawn on the first attempt, and that a coin flipped simultaneously with the marble drawing will land on heads instead of tails?
Analysis: *Choice G is correct.* The probability of drawing the marble is 1/4, and the probability of heads occurring is 1/2. These are independent events, so the probabilities combine to 1/4 x 1/2 = 1/8 = 0.125. This means that the black marble-heads combination should occur 12.5 times out of every 100 times the event is tried.

50 On the last test in Mr. Counter's math class, the median score was an 87 and the mode was 92. Harold argued that this meant the mean score must be higher than 87. Is Harold correct?
Analysis: *Choice D is correct.* Harold's assumption is incorrect. The information given is not enough to determine the mean score. Although the median and mode are relatively high numbers, you cannot be sure whether the scores beneath the median are considerably lower than the scores above the median. For example, the highest score below the median could be a 40. That would mean every other score below the median was lower than 40, which would lessen the mean considerably. The mode is also not a good indicator. Even though the mode does indicate the value appearing the most in a given set of data, the value of the mode may only appear twice in a set of 20 numbers. There is no indication that there was a large number of students who scored 92, only that more students scored 92 than any other score.

Mathematics Assessment Two: Correlation Chart

The Correlation Charts can be used by the teachers to identify areas of improvement. When students miss a question, place an "X" in the corresponding box. A column with a large number of "Xs" shows more practice is needed with that particular standard.

Correlation	MA.A.1.4.2	MA.A.1.4.2	MA.A.1.4.4	MA.A.3.4.1	MA.A.3.4.3	MA.A.3.4.3	MA.A.4.4.1	MA.B.1.4.1	MA.B.1.4.2	MA.B.1.4.3	MA.B.2.4.1	MA.C.1.4.1	MA.C.1.4.1	MA.C.1.4.1	MA.C.1.4.1	MA.C.2.4.1	MA.C.2.4.1	MA.C.3.4.1	MA.C.3.4.1	MA.C.3.4.2
Answer	C	F	D	G	*	*	A	G	*	*	B	*	H	*	*	*	B	F	*	A
Question	1	2	3	4	5	6	7	8	9	10	11	12	13	14	15	16	17	18	19	20

Student Names

* Gridded-Response Item

Mathematics Assessment Two: Correlation Chart

Correlation	MA.C.3.4.2	MA.D.1.4.1	MA.D.1.4.1	MA.D.1.4.1	MA.D.1.4.1	MA.D.1.4.2	MA.D.1.4.2	MA.D.1.4.2	MA.D.1.4.2	MA.D.1.4.2	MA.D.2.4.2	MA.D.2.4.2	MA.D.2.4.2	MA.E.1.4.1	MA.E.1.4.2	MA.E.1.4.2	MA.E.2.4.1	MA.E.2.4.1	MA.E.3.4.1	MA.D.1.4.1
Answer	I	*	C	F	B	G	*	D	F	*	B	*	*	H	*	A	*	G	*	B
Question	21	22	23	24	25	26	27	28	29	30	31	32	33	34	35	36	37	38	39	40

Student Names

* Gridded-Response Item

Mathematics Assessment Two: Correlation Chart

	MA.A.3.4.2	MA.C.2.4.1	MA.B.1.4.1	MA.B.1.4.1	MA.C.3.4.1	MA.C.3.4.1	MA.B.2.4.1	MA.B.2.4.2	MA.E.2.4.2	MA.E.3.4.2
Correlation										
Answer	H	B	H	B	H	A	H	B	G	D
Question	41	42	43	44	45	46	47	48	49	50

Student Names

* Gridded-Response Item

Notes

Notes

Notes

Notes

Notes

Show What You Know® on the FCAT
for Grade 9 — Additional Products

Student Self-Study Workbooks for Reading

Flash Cards for Mathematics and Reading

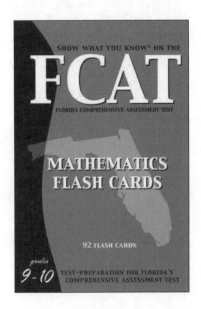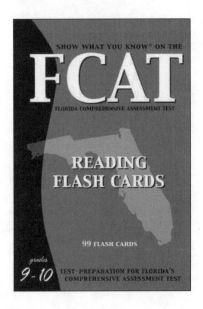